Water Services in Small Towns: Experiences from the Global South

In Focus – Special Book Series

Water Services in Small Towns: Experiences from the Global South

Editors
Klaas Schwarz and Mireia Tutusaus

IWA PUBLISHING

Published by **IWA Publishing**
Alliance House
12 Caxton Street
London SW1H 0QS, UK
Telephone: +44 (0)20 7654 5500
Fax: +44 (0)20 7654 5555
Email: publications@iwap.co.uk
Web: www.iwapublishing.com

First published 2019

British Library Cataloguing in Publication Data
A CIP catalogue record for this book is available from the British Library

ISBN: 9781789060607 (paperback)
ISBN: 9781789060614 (eBook)
ISBN: 9781789060621 (ePUB)

Contents

Water services in small towns in developing countries: at the tail end of development

Mireia Tutusaus[a,*] and Klaas Schwartz[a,b]

[a]*IHE Delft Institute for Water Education, Westvest 7, 2611AX Delft, The Netherlands*
**Corresponding author. E-mail: m.tutusaus@un-ihe.org*
[b]*Department of Human Geography, Planning and International Development Studies, Amsterdam Institute for Social Science Research, University of Amsterdam, Nieuwe Achtergracht 166, 1018 WV Amsterdam, The Netherlands*

Abstract

The current lack of knowledge about small towns, the great diversity of such settlements and the pressure to 'scale up' interventions make it difficult for policy makers and practitioners to develop models which are suitable for these towns. Currently, principles and practices informing models for water services in urban and rural areas are applied in small towns without question. This paper highlights how these principles that are engrained in the sector, may be pernicious for expanding water services in small towns as the realities of these towns may not be conducive for models incorporating these principles. Often these models are adapted to the realities of small towns to benefit water services provision. However, these adaptations are rarely documented. The little documentation of these adaptations is a consequence, we argue, of two main factors. First, such adaptations question the suitability of the models present in the sector and alter the principles on which these models are based. Secondly, the great range of adaptations linked to the diversity of small towns. Taking such adaptations seriously then limits the replicability of models, making them less suitable for scaling up interventions. As a result, certain dogmas underlying water services models continue to be reproduced.

Keywords: Development; Models; Small town; Water supply

1. Introduction

The Water services sector has long been characterized by a strong urban-rural dichotomy. The foundation for the urban-rural distinction lies in the belief that 'the urban' and 'the rural' require different service modalities in terms of technology used for supplying services, the organization and management of service provision and the financing regime for providing services. Underlying this distinction is that the urban and rural water services markets are perceived as being substantially different. The urban market is generally characterized as high population density and a relatively high level of income of

doi: 10.2166/wp.2018.001

consumers who demand a relatively high level of service. The rural market concerns an area with low population density and lower levels of economic development. Whereas urban water services would be provided through a structured network with in-house connections, rural services would be provided through either a simple network or point sources. The different types of service infrastructures also demand different capacities of the provider. In the case of urban services, an autonomous service provider operates the system. In the case of rural services community management was promoted as the main management model (Moriarty *et al.*, 2013). Over the past decades the urban water services sector has received considerable attention, particularly as a result of the heated privatization debates (Schwartz, 2008; Tutusaus *et al.*, 2018). At the same time, a distinct group of academics and practitioners focused on the rural water sector and thoroughly documented the history and development of current approaches to rural water supply (Moriarty *et al.*, 2013).

1.1. The bias towards large urban centres

Although urban water services have been subject to study over the past decades, the academics and practitioners studying the urban water services sector have largely limited themselves to studying large urban centres, whilst neglecting small urban centres. This bias towards large cities has a long tradition. Under colonial rule, 'small cities were only encouraged to expand if they were convenient administrative centers through which colonial governments could extend political control over the hinterlands or if they could serve as transfer points in transportation systems designed to exploit raw materials and agricultural resources' (Rondinelli, 1983: 43). Following this logic, development policies focused on the expansion of regions with exploitable resources and favored large urban centers, whilst largely ignoring small towns (Rondinelli, 1983; Owusu, 2005). Following Independence, the uneven development of basic water services continued as a result of the way in which governments tend to allocate resources. Gugler (as cited in Bakker, 2003: 333) highlights that governments reallocate 'resources with three goals: improving the immediate environment (of decisionmakers); assuring the continued collaboration of the middle class; and placating strategically elements of labour'. The result is that these 'public resources are disproportionately spent on the privileged consumption of the few and conspicuous investment for the few in cities'.

As a result, residents of smaller cities[1] suffer from lower levels of access to basic public services such as piped water, waste disposal, electricity and schools in comparison to residents of larger urban centers (Cohen, 2006; Ferre *et al.*, 2012)[2]. In fact, 'small urban centers in many rural regions lack the infrastructure and facilities as well as transportation and communications linkages needed to promote regional growth, social transformation and spatial integration' (Rondinelli, 1983: 379). There are specific characteristics of small towns that seem recurrent in explaining the lack of progress in these settlements. First, the cost of providing services to a widely scattered population is relatively high. Generally, small towns are characterized by relatively low population densities which hamper the realization of economies of density. As a result, centralized systems may not represent a viable service modality as the costs of such infrastructure cannot be recovered from the revenue that the

---

[1] Note that the smaller category according to this study is 100,000 inhabitants. Even though there is a mismatch in terms of population numbers in other studies used in this discussion, we could expect these trends to reflect equally disadvantageous situations in smaller urban centers.

[2] No specific data on small towns at aggregated level is available since 2006 (UN-HABITAT, 2006; Cohen, 2006).

provider can obtain from the users (Rondinelli, 1983; Adank & Tuffuor, 2013). Second, there is a presumed mismatch between the technical and financial capacity available at the local level and the requirements demanded by increasingly complex water supply systems (Njiru & Samson, 2002; Mara & Alabaster, 2008). Last, the dynamic and relatively fast growth of small towns often requires the implementation of more flexible structures rather than fixed designs of treatment plants and distribution networks. Lack of information on current developments in small towns, as well as poor planning strategies do not facilitate the effective development of infrastructure (WaterAid, 2010; Moriarty *et al.*, 2013; World Bank, 2017a, 2017b).

The urban bias is also reflected in the academic literature on water services. Much academic literature dealing with water provision in urban areas mainly depicts certain types of cities. The cities that absorb most of the attention of academics are the large economic and political centers of a country. In case other, relatively smaller, cities or towns appear in literature they are often studied in relative terms to those larger economic centers (Bunnel & Mariganti, 2010). Moreover, the literature on water services in small towns that does exist 'shows a striking degree of unanimity, indeed it seems that [...] there is a tendency for multiple authors and institutions to draw and recycle a limited number of primary sources' (Ryan & Adank, 2010: 4). Illustrative of the lack of recognition of small towns are the reductionist, yet broad, definitions used to describe such settlements. According to such definitions a small town is 'a settlement that is sufficiently large and dense to benefit from the economies of scale offered by piped systems, but too small and dispersed to be efficiently managed by a conventional urban water utility' (WSP, 2002; Pilgrim *et al.*, 2004). What these definitions do is merely highlight that small towns' water supply systems exceed the requirements of a rural approach (WaterAid, 2010), whilst also do not enjoy the scale and densities required to make a typical urban approach economically viable (Pilgrim *et al.*, 2004). In doing so, such definitions perpetuate the urban-rural dichotomy widely present in the water services sector.

2. The incompatibility of water services models with the realities of small towns

> *'A first class water works man in a small town needs the wisdom of a Solomon, the patience of a Job, the political tack of a Jim Farley, the tenacity of a General McArthur, the diplomatic ruthlessness of a Von Papen, the strength of a Pittsburg steel worker and a doctor's degree in engineering.'*
>
> Dillery (1945: 1185)

As highlighted above, small towns are frequently described as being neither urban nor rural, whilst having characteristics of both urban and rural areas. In reality, these small urban centers will typically appear to have rural socio-economic features while requiring urban-type technology for basic services (Hopkins, 2003). The different characteristics also indicate that within small towns considerable diversity may exist as to the more rural or urbanized features of specific areas within a town. More densely populated, urban-like centres may be bordered by less densely populated, rural-like areas. The absence of a clear and common definition of small towns reflects the challenge in capturing such diversity in one definition. Settlements considered small towns may have a few thousand inhabitants in some countries or up to 100,000 in others (Otiso, 2005). As such, there is no common threshold in population numbers that clearly defines a small town. The size of a city is only based on where the city limits officially cover under its administrative boundaries, but this is 'quite arbitrary' and may not include large numbers of people living adjacent to the 'proper' city (Cohen, 2006: 66). 'Levels of urbanization even in major urban areas seem to be on a continuum between purely urban and the purely rural and this point complicates the derivation of hard and fast definitions' (Choguill, 1989: 268). The diversity in size and population distribution is complemented by a diversity in economic activities that characterize the towns. Whereas some towns may be characterized by agriculture-based economies, others may have more mixed economic activities or act as regional trading centers (WaterAid, 2010).

As a consequence of the little research undertaken concerning water services in small towns, practitioners and policy makers have little information or knowledge on approaches to service provision in small towns. Instead, what we know about development of water services in small towns is understood through either 'big theories' on urban development or through well-documented approaches to rural water services. Settlements are mostly qualified and subsequently categorized as being either urban or rural (UN-HABITAT, 2006). However, working in these two apparently distinct categories blurs the reality of those people living in those areas that do not clearly fit in such distinct categorization (Cardone & Fonseca, 2006; Satterthwaite, 2006). These urban or rural approaches lack reflection on the actual characteristics and complexities of small towns, making the application of these approaches to small urban centers problematic, if not impossible (Bell & Jayne, 2009: 684). The result is that standardized models for water service provision are implemented in these areas where perhaps other approaches are required (Choguill, 1989).

2.1. Models of water service provision

Models are essentially constituted of three elements, namely, the physical infrastructure used to provide the service, the organization that manages and operates that technology, and the financial arrangements that maintain that specific service. The importance of water service models lies in the fact that such generic approaches allows practitioners and policy makers to upscale development of water services by applying a similar model in different places (Cleaver & Franks, 2008). For such models to be easily upscaled, they need to be relatively simple and sufficiently broad. Such reductionist models allow for easier transfer. Models, which have been presented or are perceived as being very successful, may then be implemented across a wide variety of locations (Rap, 2006). Despite the increased relevance of small towns, very few new models for water services have been developed in recent years, particularly in relation to small towns. This inertia for developing new models and how these have been shaped for small towns can be traced to a series of dominant discourses and principles that have become engrained in the water services sector, largely adopted by practitioners and policy-makers alike. These discourses and principles have become so dominant that they largely dictate the design of current water service models, and with it stifle any innovation. Moreover, these principles are considered applicable to water services across the rural-urban continuum and are well established almost to the point of being unquestionable, also for small towns. The incompatibility of models based on these principles is elaborated on below by highlighting somewhat interrelated factors.

2.1.1. The Promotion of commercial finance. Although different assessments show considerable variation, projections concerning the amount of funds required to finance the realization of the Sustainable Development Goals (SDGs) are significant and increasing (Winpenny, 2015). Global estimates range from US$ 6.7 trillion by 2030 to US$ 22.6 trillion by 2050. A more recent study that quantifies only the recurrent expenditure (such as operations and maintenance, staffing, or electricity) suggests that the required expenditure will increase from US$ 4.2 billion per year to US$ 31.1 billion per year between 2015 and 2030, outweighing the costs of capital investments by about 1.5 times (Hutton & Varughese, 2016). At the moment the sector is only able to cover about one-third of these needs (Development Finance, 2016).

Despite an annual increase of the government budget allocation available for water services of 4.9%, over 80% of the surveyed countries report insufficient financial means to meet national targets for drinking water and sanitation in both urban and rural areas (GLAAS, 2017). Typically, the shortcomings encountered in national budgets have been filled in by Overseas Development Aid through grants and/or loans. However, aid commitments by bilateral donors and multilateral development banks have decreased in the last years (21% from 2012 to 2015, in Sub-Saharan Africa the reduction amounts to 44%) (GLAAS, 2017). Additionally, the external support that is still received is changing its target moving away from supporting governmental budgets for water supply and sanitation to supporting projects that seek to develop 'self-reliance and sustainability'. As such, many objectives collected in donor agendas are geared to 'supporting the sustainability of WASH services', 'Strengthening policy/institutions for sustainable WASH service delivery', 'WASH financing and leveraging private sector investments', 'WASH financing and mobilizing private sector investment' or 'innovative technological solutions' (GLAAS, 2017). This highlights the efforts made in the sector to relieve national governments from responsibilities to directly provide basic services, shifting the attention to other parties. Moreover, the figures as currently collected obscure imbalances on how available funds are used for development of water services and highlight another shift in attention away from water and sanitation. When assessing allocations from external aid for 'basic WASH' services rural areas receive considerably less assistance than large urban areas (see Humphreys *et al.*, 2018). Other development priorities such as health, refugees or humanitarian assistance are being currently prioritized ahead of water and sanitation issues (GLAAS, 2017; OECD, 2017).

Over the past decades, donors have heavily promoted commercial finance as a means to bridge the financing gap. They argue that commercial finance will instil the discipline and efficiency that is required to achieve results in the sector and that public funds or operations have not been able to deliver: 'Increasing the level of private financing for the sector would allow service providers to borrow and invest in expanding service and improving quality, without having to wait for scarce public resources to be made available or rely on limited concessional financing' (World Bank, 2017a, 2017b: 24). By pursuing commercial finance, the sector also reinforces the idea that the funds required for development of water services is mainly to come from revenue generation by the water provider (cost-recovery) and less so from governmental support. It suggests that governments are not able and should not be burdened with the financing development of the water supply and sanitation sector. The dominance of this paradigm, however, may particularly impact utilities operating in small towns. These utilities, which not only have difficulties attracting public funding, are also likely to face challenges in accessing commercial finance (Humphreys & Schwartz, 2018). These utilities have often limited technical and financial capacity and are challenged in achieving creditworthy ratings as demanded by commercial lenders. As such, they do not instil sufficient confidence in private investors concerning the returns that their investment may generate. Moreover, the relative market size that the utilities serve and the limited economies of density mean that it is unlikely, or at least challenging, that the utility can operate on the basis of full cost-recovery. This makes it difficult to realize the short- to medium-term results that commercial bank lending is looking for (GLAAS, 2017).

2.1.2. Water providers as commercial businesses. Linked to the importance given to commercial finance, is the emphasis placed on of having water providers operate as business-like entities. Models that are currently endorsed in the water services sector have been heavily inspired by the privatization decade (1993–2003). Although this decade failed to lead to a proliferation of private sector organizations active in the water services sector, it did lead to the firm anchoring of private sector practices and principles within public water utilities. Public utilities, either at their own initiative or under (donor) pressures, emphasized efficiency gains, performance management, results orientation and other approaches usually associated with the private sector. This emphasis on private sector principles is not something that is limited to water utilities servicing large urban centres.

A model for water user associations developed by Water Aid (n.d.), for its implementation in semi-rural settings, promotes a management system that is based on a business principle of cost recovery and profit-making. Rusca *et al.* (2015) highlight how this model for water user associations providing water services has largely been molded to mimic private sector organizations. Similarly, the need to improve efficiency and professionalism has also been 'largely' reported on typically rural management models such as community-based arrangements (Moriarty *et al.*, 2013). As a result, water providers, regardless of the city, town or village in which they operate, are encouraged to operate as autonomous companies on the basis of full-cost recovery, be entrepreneurial and strive for ever more efficient utility operations through elaborate systems of performance management.

Water providers, thus, not only have to ensure reliable provisioning of a vital resource, but also have to address commercial objectives (Tutusaus *et al.*, 2018). Sustainability is often reduced to a perspective that largely limits itself to equate sustainability at the lowest possible level, usually perceiving the need to deliver financially viable services at the level of individual systems avoiding any interference or support from external systems. The system that is able to close the circle, mainly in financial terms, is perceived as the ultimate achievement of success and will most likely be portrayed as a 'best practice' (see Rautanen & White, 2018).

These two trends deeply engrained in the sector are of particular significance for small towns. The relatively smaller market sizes, lower population densities and economies profiles potentially clash with the 'as desired' implementation of these principles. Although we acknowledge the importance that lays on the efficient management of water services and financial sustainability, as well as the burden of overwhelmed governments we do question what these pursue of these principles does to ensuring universal water provision.

3. Special issue on water supply in small towns

Despite the moderate attention from academia, the importance of small towns is gaining increased recognition as a result of two developments. The first development concerns the possible role of small towns in migration flows and urbanization. Development of small towns, in the form of improved basic services and other amenities, has been promoted in order to abate the impact of urbanization on large urban centres and to alleviate service provision pressures in major urban centers, whilst stimulating rural economies and eventually prompting social transformations (Otiso, 2005; Bell & Jayne, 2009; Roberts, 2016). The second development concerns the targets set by the SDGs in 2015. SDG6 requires countries to ensure universal and equitable water services by 2030. This inclusive target requires that the water services needs of small towns are considered in the expansion of sustainable and equitable water services. In this Special Issue, we aim to contribute to the study of water services in small towns by critically examining different approaches and experiences of water supply in small towns. The special issue brings together empirical testimonies of how the implementation of reductionist models and the perseverance of certain principles underlying these models in the water sector have yielded sub-optimal results.

3.1. Explaining what works

In the articles of this special issue we see how the credo that private operators are synonymous of efficiency, accelerated access, increased transparency and more sustainability trickles down to different contexts in small towns of Mozambique or Nepal. In Mozambique, the government agency in charge of expanding services to small towns insists on engaging in contracts with local private operators. Despite the fact the local private sector is mostly inexistent in the country, the governmental agency in charge makes it part of their mandate to expand and capacitate these yet to be developed private sector in order to develop small-scale private operators able to operate in the context of small towns. The delegation of water operations to small-scale private operators does, however, not automatically lead to successful private operators. Rather it asks for tailored efforts from governmental agencies and for a significant reinterpretation of the contractual relationship between public (governmental agencies) and private (operators) entities (Tutusaus *et al.*, 2018). In Nepal, only one operator in semi-rural areas in the country has managed to expand services and operate efficiently. However, the case suggests that professionalization and commercialization in themselves do not fully explain the success of service expansion in semi-rural areas of Nepal (Rautanen & White, 2018). Both cases show that explaining what works needs to go beyond the reductionist explanations engraved in the models applied to most small towns. Without necessarily doing away with the focus on efficiency, professionalization and commercial operations of small-scale providers the cases show that by themselves these factors are not enough to explain why a particular model is working. Other additional factors are required to make the model work. Rather than assume small-scale providers have certain capacities, it may be more important to document what is required and possible in these settings and which demands can be made of such operators.

3.2. The heterogeneous landscape

The need for working with models leads to a necessary simplification which requires a fair amount of assumptions made about local realities. The diversity of small towns, however, may not lend itself to such broad assumptions. Humphreys & Schwartz (2018) elaborate on the role of public finance in developing water infrastructure in small towns in Burkina Faso. While national government budget allocations exist for the expansion of water services in small towns, most of these allocated funds do not reach small towns. The little local capacity to generate funds, as well as the limited capacity to claim funds from the state budget leaves towns searching for alternative funds such as project financing and donor funds. As a result, small towns resort to accessing funds from alternative sources such as donors or private sector donations. As the ability of small towns to access such funds differs, however, different small towns have varying degrees of success in accessing such funds. What results is an uneven landscape of water infrastructure development across small towns. This finding suggests that models currently promoted for small towns may be more suitable for some types of small towns, whilst other small towns may need alternative models and approaches.

3.3. Documenting adaptions

A considerable discrepancy exists between the models as they are described on paper and how they are implemented on the ground. Water services models are often adjusted to adapt to local conditions in small towns. Several papers in this issue analyse adjustments made to the policy models on paper in order to accommodate requirements and capacities in the small towns. In Adank *et al.* (2018) the modification of a framework for the evaluation of water service sustainability is presented. This framework had originally been designed for larger urban areas. During the translation of the framework to its implementation to small towns in Ethiopia, however, it was realized that not all parameters were applicable or reasonable to request from such settlements. In some instances, items that remained part of the eventual framework still do not provide an accurate picture of the actual service configurations in the towns. More often what seems to occur is that requirement adjustments are not collected in official documentation. What results is a dual system of actual operations and an associated level of performance as it is practiced, and a paper-based reality of how the model was initially designed and, by all official accounts, is still working. A striking example of this duality is the delegation of services to local private operators in Mozambique (Tutusaus *et al.*, 2018). The case suggests contracting in small towns of Mozambique is much more relational in nature than the contracts and tendering procedure suggest. Similarly, Lele *et al.* (2018) describe how their case studies in India are small towns that officially are to source their water from surface water only. However, informally the water providers in these towns access alternative sources of water, such as groundwater, in order to facilitate the expansion of services. Moreover, Molle (2008) highlight how this sourcing of alternative sources increases reliability of service provision. This practice, however, is not formally acknowledged or officially collecting which prompts the authors to advocate a revision to the model to allow for 'mixing' sources of water to meet growing demand in small towns. That projects are adapted during implementation and adjusted to local conditions has been documented at large (Molle, 2008). We would like to concur with that point and furthermore add that these adjustments are not recorded for two reasons. As we have said before, the workings of models requires simplifications and acknowledging all adaptations does not help the process of scaling up (Cleaver & Franks, 2008). Furthermore, the recording of these adjustments could translate into an official acknowledgement that those principles that underlie the model contradict the local realities of small towns and may subsequently create the space to question these very founding principles.

4. Conclusions

Underlying the challenge of models for small towns and the requirement of local adaptations is the tension between the need for simple models and the complexity and diversity of small towns. On the one hand, the development of tailor-made models for each and every small town is impossible as the number of small towns is too large and the effort required to design such customized models too laborious. Policy makers and practitioners require relatively simple models which they are able to roll out over a large number of settlements if they are to achieve the desired rates of service expansion. At the same time, these settlements show such a vast diversity that any model would require some alterations to local specificities before it can be expected to function. The articles in this Special Issue highlight that in the water services sector, and in understanding water services in small towns, we are still far removed from the goal of reconciling this tension.

In this process, the lack of understanding of the needs, challenges, as well as opportunities in small towns also does not offer a very firm foundation on the basis of which to rethink models for expanding services in these type of settlements. The majority of the examples documented in this Special Issue are a result of implementing different models through trial and error. As some of the articles show, in some cases local adaptations, mostly implemented by water practitioners, have resulted in relatively successful cases of service provision. But more often, however, such adaptations failed to materialize. The limited capacity of water providers in small towns to absorb model malfunctions means that in such cases the impact on services provided will be both clearly visible and severe for consumers.

What is even more disheartening is that many adaptations that are made are not even documented or analysed. This lack of documentation has two important consequences. First, it leads to a situation where the principles and practices underlying the models currently being implemented in small towns are hardly challenged. Although these principles may be suitable in some cases, questions can be raised about the general applicability of these principles for small towns in general. Without documentation of adaptations that deviate from the established principles, it is unlikely that the general applicability of these principles will be contested. As long as those are not contested they will continue to inform and shape the models for water service provision in small towns of the future. Secondly, the adaptations provide valuable insights into making models more resilient and, thus, less likely to fail. The multiple sources of water presented in the Indian cases or the relational contracting approach argued for involving small-scale operators in Mozambique provide examples of how models are being made more flexible to allow for more generic models to be better customized to specific situations (such as water scarcity or of limited private sector capacity).

The inertia of developing models for water services more suitable for small towns has meant that these settlements find themselves at the tail end of water supply and sanitation development. Apart from the urban bias which favours spending public funds on basic services in large metropolitan areas, the neglect of being neither urban nor rural, and that very few models for development of water supply and sanitation have been developed specifically for small towns means that development of water service in such settlements lags behind. Although the SDG target to achieve universal service coverage by 2030 may provide a strong impetus to start taking the tail end of water infrastructure development seriously, much remains to be done before achieving universal service coverage in small towns is likely. In order to do that, we should start speaking of small towns as a category on their own and continue the work in elaborating further what these are and how they work.

References

Adank, M. & Tuffuor, B. (2013). *Management Models for the Provision of Small Town and Peri-Urban Water Services in Ghana. TPP Synthesis Report*. WASH Resource Center (RCN). Accra, Ghana.

Adank, M., Godfrey, S., Butterworth, J. & Defere, E. (2018). Small town water services sustainability checks: development and application in Ethiopia. *Water Policy 20*(s1), 52–68.

Bakker, K. (2003). Archipelagos and networks: urbanization and water privatization in the South. *The Geographical Journal 169*(4), 328–341.

Bell, D. & Jayne, M. (2009). Small cities? Towards a research agenda. *International Journal of Urban and Regional Research 33*(3), 683–699.

Bunnel, T. & Mariganti, A. (2010). Practising urban and regional research beyond metrocentricity. *International Journal of Urban and Regional Development 34*(2), 415–420.

Cardone, R. & Fonseca, C. (2006). *Water Sector in Small Urban Centres. UN-HABITAT Report on the Water Sector in Small Centers: Analysis of Donor Flows to Water Supply and Sanitation Services Paper 2*. United Nations Human Settlements Programme (UN-HABITAT), Nairobi, Kenya.

Choguill, C. L. (1989). Small towns and development: a tale from two countries. *Urban Studies 26*, 267–274.

Cleaver, D. & Franks, T. (2008). Distilling or diluting? Negotiating the water research-policy interface. *Water Alternatives 1*(1), 157.

Cohen, B. (2006). Urbanization in developing countries: current trends, future projects and key challenges for sustainability. *Technology and Society 28*, 63–80.

Development Finance (2016). How to plug the gap in water investments. Issue: Infrastructure Finance. Alliance 4 Water. April 2016. http://alliance4water.org/resources/Development-Finance-03.pdf (accessed 29 November 2017).

Dillery, D. R. (1945). Problems of the small town water works. *Journal American Water Works Association 37*(11), 1185–1190.

Ferre, C., Ferreira, F. & Lanjouw, P. (2012). Is there a metropolitan bias? The relationship between poverty and city size in a selection of developing countries. *The World Bank Economic Review 26*(3), 351–382.

GLAAS (2017). *Financing Universal Water, Sanitation and Hygiene Under the Sustainable Development Goals: UN-Water Global Analysis and Assessment of Sanitation and Drinking Water. GLAAS 2017 Report*. World Health Organization, Geneva.

Hopkins, R. (2003). *An Alternative Perspective on WSS Services (Including the 'Grey Area'). Town Water Supply and Sanitation Services. Vol. 3. Companion Paper A1. Bank-Netherlands Water Partnership Project num. 43*. The World Bank. Washington, DC.

Humphreys, E. & Schwartz, K. (2018). In the shadow of the city: financing water infrastructure in small towns in Burkina Faso. *Water Policy 20*(s1), 69–83.

Humphreys, E., van der Kerk, A. & Fonseca, C. (2018). Public finance for water infrastructure development and its practical challenges for small towns. *Water Policy 20*(s1), 100–111.

Hutton, G. & Varughese, M. C. (2016). *The Costs of Meeting the 2030 Sustainable Development Goal Targets on Drinking Water, Sanitation, and Hygiene. Water and Sanitation Program Technical Paper*. World Bank, Washington, DC. Available at: http://www.worldbank.org/en/topic/water/publication/the-costs-of-meeting-the-2030-sustainable-development-goal-targets-on-drinking-water-sanitation-and-hygiene (accessed 29 November 2017).

Lele, S., Madhyastha, K., Sulanga, S., Dhavamani, R. & Srinivasan, V. (2018). Match, don't mix: implications of institutional and technical service modalities for water governance outcomes in south Indian small towns. *Water Policy 20*(s1), 12–35.

Mara, D. & Alabaster, G. (2008). A new paradigm for low-cost urban water supplies and sanitation in developing countries. *Water Policy 10*, 119–129.

Molle, F. (2008). Nirvana concepts, narratives and policy models: Insights from the water sector. *Water Alternatives 1*(1), 131–156.

Moriarty, P., Smits, S., Butterworth, J. & Franceys, R. (2013). Trends in rural water supply: towards a service delivery approach. *Water Alternatives 6*(3), 329–349.

Njiru, C. & Sansom, K. (2002). Managing watsan services in small towns. In: *Proceedings of the 28th WEDC Conference*, Calcutta, India.

OECD (2017). Roundtable on Financing Water. Session 4: Background note Blended finance for water security investments. Tel Aviv, 13 September 2017. http://www.oecd.org/environment/resources/Session%204%20Blended%20finance%20for%20water%20security%20investments.pdf (accessed 29 November 2017).

Otiso, K. (2005). Kenya's secondary cities growth strategy at a crossroads: which way forward? *Geography Journal 62*(1), 117–128.

Owusu, G. (2005). Small towns in Ghana: justifications for their promotion under Ghana's decentralisation program. *African Studies Quarterly 8*(2), 48–69.

Pilgrim, N., Roche, B., Revels, C., Kingdom, B. & Kalbermatten, J. (2004). *Town Water Supply and Sanitation. Bank-Netherlands Water Partnership Project No. 43*. The World Bank, Washington, DC.

Rap, E. (2006). The success of a policy model: irrigation management transfer in Mexico. *The Journal of Development Studies 42*(8), 1301–1324.

Rautanen, S. & White, P. (2018). Portrait of a successful small-town water service provider in Nepal's changing landscape. *Water Policy 20*(s1), 84–99.

Roberts, B. H. (2016). *Managing Systems of Secondary Cities: Policy Responses in International Development.* Cities Alliance. Brussels, Belgium.

Rondinelli, D. (1983). Dynamics of growth in secondary cities in developing countries. *Gegraphical Review 73*(1), 42–57.

Rusca, M., Schwartz, K., Hadzovic, L. & Ahlers, R. (2015). Adapting generic models through bricolage: elite capture of water users' associations in peri-urban Lilongwe. *European Journal of Development Research 27*(5), 777–792.

Ryan, P. & Adank, M. (2010). *Global Best Practice in the Management of Small Town Water Supplies.* IRC International Water and Sanitation Centre, The Hague, The Netherlands.

Satterthwaite, D. (2006). *Outside the Large Cities: the Demographic Importance of Small Urban Centers and Large Villages in Africa, Asia and Latin America. Human Settlements Discussion Paper – Urban Change 3.* London.

Schwartz, K. (2008). The new public management: the future for reforms in the African water supply and sanitation sector? *Utilities Policy 16*(1), 49–58.

Tutusaus, M., Schwartz, K. & Smit, S. (2018). The ambiguity of innovation drivers: the adoption of information and communication technologies by public water utilities. *Journal of Cleaner Production 171*, S79–S85.

Tutusaus, M., Cardoso, P. & Vonk, J. (2018). (de)Constructing the conditions for private sector involvement in small towns' water supply systems in Mozambique: policy implications. *Water Policy 20*(s1), 36–51.

UN-Habitat (2006). *Meeting Development Goals in Small Urban Centres – Water and Sanitation in the World's Cities.* Earthscan.

WaterAid (2010). *Small Town Water and Sanitation Delivery: Taking a Wider View.* WaterAid, UK.

Winpenny, J. (2015). *Water: fit to Finance? Catalyzing National Growth Through Investment in Water Security, Report of the High-Level Panel on Financing Infrastructure for a Water-Secure World.* World Water Council and OECD.

World Bank (2017a). *Sustainability Assessment of Rural Water Service Delivery Models: Findings of a Multi-Factor Country Review. Water Global Practice. August 2017.* World Bank, Washington, DC.

World Bank (2017b). *Sanitation and Water for All: How can the Financing Gap be Filled? A Discussion Paper.* World Bank, Washington, DC. https://openknowledge.worldbank.org/bitstream/handle/10986/26458/114545-WP-P157523-PUBLIC-SWA-Country-Preparatory-Process-Discussion-Paper-8-Mar-17.pdf?sequence=1&isAllowed=y (accessed 19 November 2017).

WSP (2002). Volume 1: Summary of International Conference on Water Supply and Sanitation for Small Towns and Multi-Village Schemes, Proceedings of the International Conference, Addis Ababa, Ethiopia, June 11–15, 2002.

Match, don't mix: implications of institutional and technical service modalities for water governance outcomes in south Indian small towns

Sharachchandra Lele*, Karthik Madhyastha, Swati Sulagna, R. Dhavamani and Veena Srinivasan

Centre for Environment and Development, Ashoka Trust for Research in Ecology and the Environment (ATREE), Royal Enclave, Srirampura, Jakkur P.O., Bengaluru 560064, India

**Corresponding author. E-mail: slele@atree.org*

Abstract

This paper seeks to contribute to the limited literature on water governance in small towns in India. For assessing water governance, we propose a broad framework encompassing adequacy and affordability, equity, sustainability and responsiveness. Analytically, the concept of 'service modality' is expanded to include not only institutional arrangements but also water resource deployment, and placed within a framework that includes multiple contextual variables as well. We use this framework to carry out an inductive analysis by comparing water service delivery and governance in four small towns across two states (Karnataka and Tamil Nadu) in southern India. Apart from differences in size, the towns differ in the institutional arrangements – from fully municipal management to a combination to complete para-statal management – and in the deployment of water resources – only ground water to a mixed supply of ground and surface water (dual sourcing). Data were gathered using a combination of household surveys, metering, records, and interviews. Dual sourcing resulted in adequate supply and optimization vis-à-vis end uses. Inter-household inequity is driven by socio-economic differences amongst households, but can be mitigated to an extent by increasing public tap density. But water resource use is not physically or financially sustainable. The responsiveness to citizen needs was significantly higher when the distribution was done by the local governments. Separation of roles, with para-statals providing bulk supply of surface water, and local governments managing the distribution of this and groundwater, may be an optimal service modality.

Keywords: Dual-sourcing; Small towns; Urban water; Water governance; Water service delivery; Water service modalities

Introduction

Urban domestic water demand is growing dramatically in developing countries because of the twin phenomena of population growth and rural–urban migration. Within the urban landscape, smaller

doi: 10.2166/wp.2018.002

towns, i.e., towns with population less than 100,000, are likely to become the centres of fastest population growth (Gupta, 2012). At the same time, such towns in developing countries face special challenges in meeting the water needs of their citizens, because of limited human and financial resources, absence of dedicated water utilities, and limited physical resources, often resulting in high dependence on groundwater (Mugabi & Njiru, 2006; Shah, 2013).

While there is a thriving literature on urban water governance in large cities of developing countries around the world, research on water supply and water governance in small towns in developing countries is of recent origin and somewhat limited in coverage and scope. Analytically, this research has highlighted the limited penetration of state-run piped water supply, especially in Africa (e.g., Adank *et al.*, 2011, p. 16) and the Philippines (Tungpalan, 2009). Consequently, the focus has been on documenting and evaluating alternative modes of water service delivery, including community-managed institutions, private players, and their combinations (e.g., Cain & Mulenga, 2009; Fragano, 2001). The Indian context, and particularly that of peninsular India, differs significantly from this in that state-run water supply systems have high levels of penetration even in small towns. So the 'service modalities'[1] debate in Indian small towns is less about how to integrate public utilities with private or community players, and more about the role of and interface between different public agencies or levels of government (Krishnan, 2006; De Bercegol & Gowda, 2013)[2]. Similarly, the global literature on water supply in small towns appears to assume that lack of infrastructure and management is the problem, not physical availability of water (see, for example, the review by Mugabi & Njiru, 2006). In the Indian context, however, where population densities are very high and water much scarcer than in some other countries, this assumptions may not hold, and the choice of whether and how to use surface water (SW) or ground water (GW) is likely to be a crucial element of water management decisions at both household and municipal levels (Shah, 2014). There is an emerging literature on the conjunctive use of SW and GW by households in cities (e.g., Srinivasan, 2008), but not much attention has been paid to 'dual-sourcing' by the water supply agencies themselves.

Normatively, the literature on small town water services has tended to focus on adequacy and access, quality, affordability, and cost recovery (e.g., Caplan & Harvey, 2010). We believe that, in addition these concerns, the question of equity in distribution, physical sustainability, and the responsiveness and accountability of the public agencies involved are also important to policy makers. The literature on these dimensions, however, is much more limited. For instance, while there is literature on physical sustainability of water management systems in villages (e.g., Rautanen & White, 2013) and there is a debate about whether large cities should be importing so much water from faraway dams (e.g., Mukherjee *et al.*, 2010), there is little discussion of groundwater sustainability, even in the context of small towns that are often highly groundwater dependent (Raju *et al.*, 2004; Shah, 2013). Similarly, a significant literature on equity in water service delivery exists for big cities (Bakker *et al.*, 2008; Anand, 2011; Subbaraman *et al.*, 2013), but hardly any on similar issues in small towns.

Moreover, there appears to be a degree of disciplinary fragmentation, with sociologists focusing on equity issues and economists focusing on efficiency, preventing an integrated understanding of the multiple dimensions of concern highlighted above. Not surprisingly then, the relationship between

---

[1] A term coined recently: Rusca & Schwartz (2012).

[2] One major debate about institutional arrangements in cities in India and elsewhere, viz., the question of privatization of water service delivery (McKenzie & Ray, 2009; Sangameswaran *et al.*, 2008) appears to not be relevant to small towns because they are too small to attract private players (Mugabi & Njiru, 2006).

institutional arrangements, choice of resource, administrative and political cultures and the socio-economic context of households on the one hand and these multi-dimensional outcomes on the other hand remains poorly explored. Such a broader framing and interdisciplinary data are essential when policy debates do emerge, such as whether municipal supply should be handed over to a specialised para-statal agency or not (Jayaramu *et al.*, 2015), whether metering is essential for cost-recovery or avoiding wastage in consumption (Kumpel *et al.*, 2017), or whether importing SW is essential or not (Nadhamuni, 2012). But this broader approach has been rare in small town water research.

In a modest attempt to fill these gaps, we present here the findings from a comparative analysis of domestic water outcomes[3] in four small towns from the states of Karnataka and Tamil Nadu in southern India that vary in the role played by public agencies and in the availability of SW to supplement GW. This study was part of a larger interdisciplinary research project carried out in two river basins in southern India; the choice of small towns was therefore confined to these basins. The study seeks to contribute at two levels. Firstly, it seeks to assess the variations in the 'outcomes' of water governance across the multiple dimensions mentioned above, viz., adequacy (including affordability and quality), equity, resource and financial sustainability, and responsiveness and accountability in water governance. Secondly, the study seeks to understand how differences in socio-technical arrangements or 'service modalities'[4], specifically in the roles of the local government versus provincial para-statal agencies and in the use of GW versus SW, may influence outcomes across these multiple dimensions. Drawing upon data gathered from household surveys, interviews, physical measurement of borewell pumping rates, and secondary data, we seek to understand what might be the appropriate role for different state agencies and how different sources of water might be optimally used, while at the same time pointing to the influence of other factors outside the service modality. We adopt an inductive approach, and so the study concludes with hypotheses rather than firm assertions, some relevant to small towns and others to urban water governance as a whole.

Following from the inductive approach, the paper begins with a description of the towns studied and an overview of the service modalities they have adopted. This is followed by a section outlining the multiple dimensions of outcomes we focused on and the framework for analysis. The section on methods provides details of the indicators used for the outcomes and how data were gathered for assessing them and for exploring potential links with the factors in the framework. We then present the results along the multiple dimensions and their relationship with intermediate decisions made by the service provider. The links with service modality and other factors are discussed in the subsequent section. In conclusion, we offer some hypotheses regarding small town water governance in India for further exploration.

Study sites and service modalities

This research was part of a larger study on climate change adaptation in the water sector in two river basins: the Arkavathy River in Karnataka state and the Noyyal River in Tamil Nadu state (see www.atree.org/accuwa for more details). Both basins are similar in having a dry climate (~600–900 mm

[3] Although use of water in commercial activities was not entirely negligible in some towns, we have not included that in this paper for want of adequate data and benchmarks as to what constitutes reasonable use. So for this paper, small town water governance is equated with small town domestic water governance.

[4] We define 'service modality' broadly to include both the institutional arrangements and sources of water deployed.

annual rainfall), hard rock aquifers at that are characterised by low water yields, a largely (75–80%) urban population, and rivers that have either been depleted or polluted. Whereas the Arkavathy basin is dominated by the metropolis of Bengaluru and has only four small towns, the Noyyal basin has the two cities of Coimbatore and Tiruppur alongside 86 smaller urban settlements.

While the choice of towns was constrained by the basin boundaries in the larger research project, there were some obvious differences in institutional arrangements, water resource choices, and town sizes. We therefore chose a sample that varied in institutional arrangement, water resource deployment, and size. For this paper, we focus on the towns of Nelamangala and Ramanagara in the Arkavathy basin, and Kannampalayam and Palladam in the Noyyal basin. The basic features of these towns are given in Table 1.

The town populations range from ~15,000 to ~100,000, thereby also capturing the two kinds of urban local bodies (ULBs) that exist in Indian small towns: Town Panchayat/Town Municipal Council for the smaller towns and Municipality/City Municipal Council for the bigger ones. All towns have experienced high population growth rates (ranging from 20% to 47% growth in a decade), have an economy based partly on small-scale textile related operations, and have similar (~15%) fractions of socially marginalised groups in their population.

The differences in service modalities are given in bold in Table 1. In terms of sources of water used, three of the towns use a combination of imported river[5] water and groundwater, whereas one (Nelamangala) is entirely groundwater dependent. In terms of the institutional arrangements for water service delivery: the towns in Tamil Nadu (Kannampalayam and Palladam) receive bulk supply of river water from a provincial para-statal agency, Tamil Nadu Water and Drainage Board (TWAD Board), but manage the distribution of this river water and of GW on their own. In contrast, in Karnataka, the arrangements vary across towns. Ramanagara has both bulk supply and distribution being done by the equivalent provincial[6] para-statal, viz., Karnataka Urban Water Supply and Drainage Board (KUWSDB). Nelamangala, however, manages its water supply and distribution entirely on its own. The combination of institutional arrangement and source of water gives three distinct service modalities:

- ULB alone, supplying only ground water,
- ULB and para-statal, supplying both ground and surface (river) water, and
- Para-statal alone, supplying both ground and surface water.

Note that there is limited self-supply by individual households in all the towns studied. Less than 5% of the households in all towns had private wells[7]. Private tanker use was seasonally significant in the two towns in the Arkavathy basin, yet most of the tanker supply was also by the ULB/para-statal.

There are additional nuances in the socio-technical arrangements for water service delivery that are common to the four towns. First, the main forms of water distribution are 'public standpipes' (street taps) and 'in-house connections' (see Figure 1). In periods of acute shortages, tankers may be deployed

[5] We use the term 'imported' to indicate that although each of the towns is on the bank of a river, the river water that they get is from another river basin through pipelines running for 50–100 km.

[6] We use the adjective 'provincial' here, because in theory the ULB itself is part of the 'state', and it could set up a dedicated water supply agency as well. In India, however, all such agencies are set up by and answerable to the provincial (state) government (Perret et al., 2014). For the rest of the paper, we refer to these agencies as 'para-statals'.

[7] These were owned by better-off households and used as supplements to town supply, and as such do not confound our findings about inequity in town supply.

Table 1. Basic features of towns studied and key elements of their service modality (in bold).

Town	Nelamangala	Ramanagara	Kannampalayam	Palladam
District (and state) located in	Bengaluru Rural (Karnataka)	Ramanagara (Karnataka)	Coimbatore (Tamil Nadu)	Tiruppur (Tamil Nadu)
River basin	Arkavathy	Arkavathy	Noyyal	Noyyal
Population in 2011 (and change since 2001)	37,232 (+47%)	95,167 (+20%)	15,868 (+33%)	42,225 (+41%)
Key economic activities	Textiles (small scale weaving)	Silk reeling (small scale)	Textiles (spinning mills)	Textiles (power looms), Poultry
Civic status of the ULB	Town Municipal Council (TMC)	City Municipal Council (CMC)	Town Panchayat (TP)	Municipality
Water source	**GW only**	**River + GW**	**River + GW**	**River + GW**
Bulk water supply by	**Not applicable (no river water import)**	**Para-statal (KUWSDB)**	**Para-statal (TWAD Board)**	**Para-statal (TWAD Board)**
Distribution by	**ULB**	**Para-statal (KUWSDB)**	**ULB**	**ULB**
Number of municipal wells	70 borewells	200 borewells (estimated)	17 open and 19 borewells	16 open and 88 borewells
Number of public standpipes (GW)	150–200 (estimated)	430	6 hand pump + public taps	171
Number of public standpipes (river water)	0	0	89	32

Note: ULB = urban local body (elected local government). TMCs and CMCs in Karnataka are equivalent to TPs and Municipalities in Tamil Nadu, respectively.

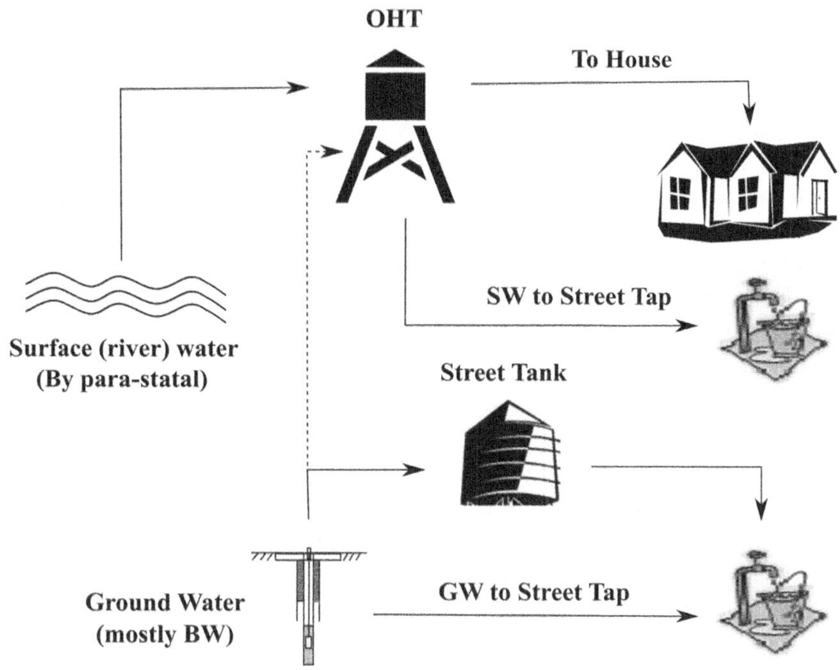

Fig. 1. Schematic diagram indicating typical public water distribution arrangements (dotted lines indicate less common pathways) (GW = ground water, BW = borewell, SW = surface water, OHT = overhead tank).

by the ULB to augment supplies. Second, in terms of frequency of distribution, in all towns, the supply is intermittent: a few hours on any given 'supply' day. Third, in terms of charging for water supply, the supply at the public standpipes is free of charge, whereas in-house connections are charged at a fixed (unmetered) monthly rate.

The administrative and political cultures in the two states are, however, somewhat different. Tamil Nadu inherited the administration set up by the British in erstwhile Madras Presidency, which was known to be efficiently run, whereas Karnataka state amalgamated administrations from different presidencies and princely states, and has been consistently ranked lower than Tamil Nadu in administrative efficiency (Mundle *et al.*, 2016). Politically, Tamil Nadu has effectively been a two-party system, whereas Karnataka has seen more fragmentation and permutation in party politics.

Framework

In policy-making in the water sector, several normative concerns are always implicitly or explicitly present, and they determine how 'performance' of water service delivery or 'outcomes' in general are assessed. For instance, the concept of integrated water management is defined by some as 'maximising economic and social welfare in an equitable manner without compromising the sustainability of ecosystems' (GWP TAC, 2000); it also refers to stakeholder participation as a process goal. Similarly, the Sustainable Development Goal 6 for water highlights *inter alia* the need for sustainability, water use efficiency and stakeholder participation. Individual academic studies, being typically driven by disciplinary perspectives, tend to focus on fewer dimensions, with the result that different definitions of 'outcome' or 'performance' result in different explanations about what 'works'[8]. We therefore believe that a multi-dimensional assessment of outcomes better reflects policy and societal concerns in this sector and allows analysis to engage with them simultaneously. Drawing upon previous analyses (Joy *et al.*, 2004; Pahl-Wostl *et al.*, 2012; Lele *et al.*, 2013), we chose the following normative dimensions and indicators along which to assess outcomes related to domestic water in the small towns we studied:

1. Adequacy, quality and affordability,
2. Inter-town, inter-ward and inter-household equity,
3. Biophysical sustainability of groundwater resource and SW storage structures,
4. Financial sustainability, and
5. Responsiveness and accountability[9].

The question then is what might shape these outcomes. Our framework for answering this question is outlined in Figure 2. This framework is tailored to the south Indian small town context of largely public supply of water, and draws upon existing debates about the role of different factors in shaping water service delivery and water governance as a whole (Krishnan, 2006; Mugabi & Njiru, 2006; De Bercegol & Gowda, 2013). The framework indicates that the outcomes are likely to be the result of multiple factors, in which decisions about water tariffs, frequency of supply, or number and location of public taps, etc., would be the proximate drivers, and amongst deeper factors, the institutional arrangements (that determine which agency is responsible for what) and water resources deployed (whether only imported SW or local groundwater or both) could play a significant role. At the same time, other factors such as individual household endowments or status, resource endowments of the town and the political and administrative cultures of the region are also likely to interact with the service modality to influence outcomes. Since ours is an inductive approach, the exact nature of relationships is expected to emerge as hypotheses from the study[10].

Methods

The above outcomes were assessed using multiple indicators that were estimated through a variety of methods and data sources, with most of the data pertaining to the year 2014–15. These indicators and the methods and data sources used are summarised in Table 2.

For understanding adequacy and equity, our original intent was to estimate average town- and ward-level water delivery in litres per capita per day (LPCD) from secondary data and then to estimate inter-household variations using qualitative methods. However, field conditions complicated matters. SW supply to the whole town could be estimated from secondary data in all three cases. But estimating ward level SW supply was only possible where the service area of the public overhead tank (OHT) that was used to

[8] Such as whether private ownership of utilities matters or not (Bakker *et al.*, 2008).

[9] It may be argued that responsiveness or accountability (elements of good governance) are means to the end of adequate, equitable or sustainable water supply. We believe, however, they are somewhat independent of the other outcomes, and that in a democratic society, these are best treated as ends in themselves.

[10] Unlike in a deductive approach, where an *a priori* theory is expected to generate hypotheses about the nature of specific links.

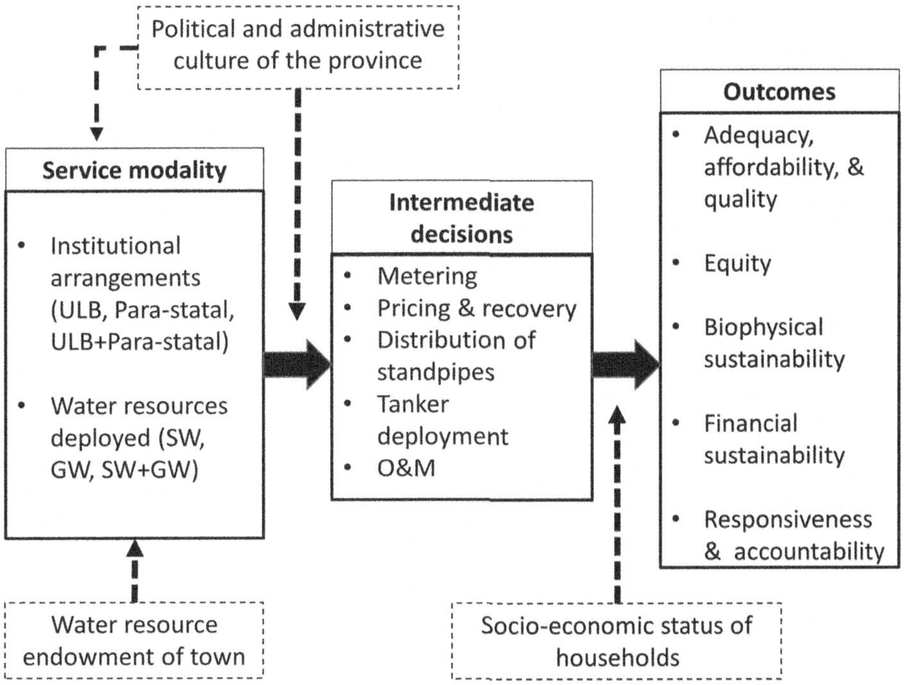

Fig. 2. Framework linking service modality and other factors to urban domestic water outcomes.

store SW matched well with ward boundaries, which was true in the Tamil Nadu towns. Estimation of the extraction of GW for public supply (see Figure 1) had never been done before, so we had to devise our own methods. For the Tamil Nadu towns, electricity consumption data were available for each of the public GW pumps (whether borewell or open well). We coupled these with pump discharge estimates by measuring discharge rates for a sample of pumps of different horsepower in each town, using either a TDS-100H ultrasonic flow meter or measuring time taken to fill the ULB's street-level tank into which the GW was being pumped. For the Karnataka towns, however, pump-wise electricity consumption data were not available from the ULB, nor did public OHT service areas correspond well with ward boundaries. Thus, only town-level estimates of public GW pumping were possible for Nelamangala and Ramanagara, by using public OHT capacities and filling frequencies obtained from linemen interviews.

To understand adequacy in the form of frequency of water supply and its seasonal variation, we used data from interviews (see below) to get at the frequency of SW and GW supply. Data on affordability came from municipal records on water tariffs, supplemented by semi-structured interviews with randomly chosen households (more than 20 in each town). These interviews also covered questions of water quality, supply frequency, and responsiveness of the administration to their complaints or concerns. In addition, semi-structured or open-ended interviews were conducted with linemen (at least five in each town), mid-level and senior ULB staff including the chief executive officers, elected town councillors (at least five in each town), and senior engineers of the para-statal agencies to understand their perceptions and constraints, and to assess their responsiveness to and involvement of citizens in planning.

For understanding inter-household variation, we focused on Nelamangala and Ramanagara. We conducted large sample surveys (N = 300 for Nelamangala and N = 600 for Ramanagara) on source-mix and water storage assets (presence/absence of sumps and private OHTs) with monitoring using water

Table 2. Methods used for obtaining data and estimating outcome variables in studied towns.

Town → Outcome ↓	Indicators used ↓	Nelamangala	Ramanagara	Kannampalayam	Palladam
Adequacy	Average water supply at town level (in LPCD)	Top-down method at ward/ OHT level (OHT volume, population served, frequency of filling of OHTs)	Top-down method at town level (total river water supplied, total town population)	Top-down method: bulk SW supply from secondary data, divided by ward populations; GW supply by monitoring of pump yield, well-wise electricity bill data and population covered	Top-down method: bulk SW supply from secondary data, divided by ward populations; GW supply by monitoring of pump yield, well-wise electricity bill data and population covered
	Frequency of water supply & seasonal variation	Field interviews and observations	Field interviews and observations	Field interviews and observations	Field interviews and observations
Affordability	Rs/kL paid; time spent	Secondary data and interviews	Secondary data and interviews	Secondary data and interviews	Secondary data and interviews
Quality	Hardness & smell	Field interviews, observation	Field interviews, observation	Field interviews, observation	Field interviews, observation
Inter-town equity	Compare average water supply across towns	From adequacy estimates above	From adequacy estimates above	From adequacy estimates above	From adequacy estimates above
Inter-ward equity	Average water supply at ward level (in LPCD)	Secondary data on OHT-level supply, source mix survey (N = 300, 11 wards), field interviews	Source mix survey (N = 600, 16 wards), field interviews	Secondary data on ward-wise supply, interviews with councillors, field interviews	Secondary data on ward-wise supply, interviews with councillors, field interviews
Inter-household equity	Average supply at household level (in LPCD)	Source mix survey (for sump ownership); Volumetric metering at hh level (N = 5) + pot survey in slums	Source mix survey (for sump ownership); Volumetric metering at household level (N = 8)	Unstructured household interviews	Unstructured household interviews
Resource sustainability	Well failure rates, water table depth, condition of SW harvesting structures	Secondary data on borewell failure and new wells dug in recent years, field interviews and observations.			
Financial sustainability	% of O&M costs covered by water charges	Secondary data on income and expenses for water supply services	Secondary data on income and expenses for water supply services	Secondary data on income and expenses for water supply services	Secondary data on income and expenses for water supply services
Responsiveness and accountability	*Various: see text*	Interviews with city councillors (N = 6), ULB staff, source mix survey	Interviews with city councillors (N = 10), para-statal engineers (N = 3), source mix survey	Interviews with councillors, ULB staff, households; secondary data from ULB	Interviews with councillors, ULB staff, households; secondary data from ULB

meters for small samples. Getting households to agree to metering of their consumption and installing the necessary meters was no easy task – we eventually managed to do this only for five households in Nelamangala and eight households in Ramanagara[11]. In addition, for households that were entirely dependent on public standpipes, we conducted a simple questionnaire survey about 'pots of water collected and used for different activities' to estimate water consumption by such households in one town (Nelamangala).

Biophysical sustainability was assessed in terms of whether GW levels are steady or declining, and the ability of local SW storage structures to harvest rainfall. Data on biophysical sustainability came from interviews with linemen, secondary information on well failure and new wells dug, and our own observations.

Financial sustainability was assessed in terms of the fraction of operation and maintenance (O&M) costs that can be met from water charges received from users. Data for calculating financial sustainability came from municipal records and senior engineer/officer interviews.

Finally, the concepts of responsiveness and accountability are part of the idea of democratic governance, but are difficult to articulate and assess. We used some basic indicators. For responsiveness, we asked how well the agencies responded to citizens' day-to-day complaints, how they responded to summer shortages, and how do they respond to scarcity in the long run. In terms of accountability, we looked at the quality of record-keeping and the transparency and participation in the longer-term planning process. Data on these aspects came from the interviews with linemen, ULB and para-statal staff, and elected representatives.

Findings

We present the findings in terms of the outcome variables identified above, and the links with the 'intermediate decisions' indicated in Figure 2, in this section. The relationship with service modalities and other factors is discussed in the subsequent section.

Adequacy

As indicated in Table 2, we used three indicators to understand adequacy: average supply (in LPCD) for the town as a whole, frequency of water supply, and change in this frequency in the summer season. The estimate of average water supplied in each town (during 2014–15) is given in Figure 3, along with the supply norms as notified by the state governments for towns of those particular size: 70 lpcd for Nelamangala and Kannampalayam, 90 lpcd for Palladam and 135 lpcd for Ramanagara (Mathur *et al.*, 2007).

There is of course uncertainty associated with these estimates, but it is not easy to quantify, because the systematic error (possible error in the estimation method) is likely to be much higher than the statistical error (variation across months or years). SW supply estimates come from secondary data, for which ULB staff gave an uncertainty of $\pm 5\%$. But groundwater supply estimates come from our

---

[11] Intermittent supply conditions required that meters be in place for at least 2 weeks to get a reasonable estimate of inflows. This made the metering-based data collection approach inherently slow. Moreover, many households that had agreed to metering, eventually removed the meters after a few days, citing 'meter reduces our inflow' as the reason. Hence the small sample size.

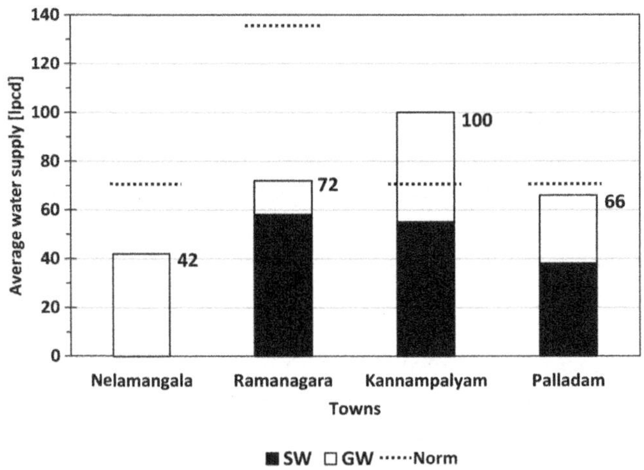

Fig. 3. Average amount of water supplied per capita in 2014–15. Numbers above the bars indicate the actual average supply. Dotted lines indicate the state's supply norm for towns of that particular size. Errors bars not estimable (see footnote 12).

one-time measurements at supply borewells and the error could as much as $\pm 20\%$. The norms should also not be applied blindly: the 135 lpcd norm is for towns with underground sewerage systems and relates to water required to move sewage, rather than water required for household needs. So it seems more appropriate to use a norm of 70–90 lpcd.

In spite of the uncertainty, and even using a modified norm, we may conclude that only Kannampalayam town meets or exceeds the official norm. Palladam and Ramanagara, however, are close to the lower end (70 lpcd) of the norm, whereas Nelamangala does not meet this norm at all. Looking at the mix of supply, it seems clear that the availability of river water is what enables higher supply levels in the three towns other than Nelamangala; groundwater alone can contribute only about 40 lpcd in such regions.

The frequency of water supply and seasonal changes in this frequency are given in Table 3. Where there is dual supply of SW and GW to each street, as in the Tamil Nadu towns, it is important to read the frequencies jointly. Keeping this in mind, we see that Nelamangala town lags behind both during normal months and in summer months, and supply frequency in Ramanagara is also not great (once in 4 days even in normal months), even though average supply in Ramanagara is comparable with that in Palladam.

The systematic 'dual-sourcing' model in the Tamil Nadu towns, wherein GW (poorer quality, because of hardness) is supplied daily or on alternate days in street taps which are spaced very close (one tap for

Table 3. Frequency of water supply during normal and summer months.

Town	Nelamangala	Ramanagara	Kannampalayam	Palladam
Supply frequency in normal months				
Surface water (SW)	Not applicable	Once in 4 days	Once in 7–10 days	Once in 5–7 days
Ground water (GW)	Once in 3 days	Not separable from SW	Daily or alternate days	Daily or alternate days
Supply frequency in summer months				
Surface water (SW)	Not applicable	Once in 8 days	Once in 10–15 days	Once in 7–10 days
Ground water (GW)	Once in 9 days	Not separable from SW	Once in 3 days	Once in 3–5 days

four dwellings or less) while SW (better quality) is supplied much less frequently, results in 'dual-use' also: households use SW for only cooking and drinking, and GW for everything else. Whereas the erratic and spatially separated model in Ramanagara, where GW supply is reserved only for areas poorly served by SW, means that no such 'dual-use' adaptation is possible.

All towns have to reduce the frequency of supply significantly during the summer, showing that although average supply during the normal months may meet the norms, significant scarcity is likely to be experienced during the lean season. The coping mechanisms adopted are also indicators of how water service delivery varies: in both the Karnataka towns, the fraction of households resorting to *occasional* (summer) purchase of water from private tankers to augment municipal water was 30–35%. But in the Tamil Nadu towns, households did not report such tanker water purchase, but relied on tanker supply by the ULB to augment supply, especially when borewells failed[12].

In terms of quality, since we relied on perception data alone, the main dimension assessed was hardness. Our interviews (not tabulated here) indicated that citizens in the Tamil Nadu towns were by and large satisfied with water quality, whereas those in the Karnataka towns were not. The reason given by the former was that they had access to better quality river water for their drinking and cooking needs. The reason for dissatisfaction in Nelamangala was obvious – they only received groundwater. Paradoxically, however, several households in Ramanagara complained that river water tended to be contaminated, and so they preferred to collect some groundwater as well. We were, however, unable to confirm the contamination or pinpoint its source.

In terms of affordability, the cost at which water is supplied varies from 0 (for public standpipes) to a fixed charge for in-house connections that ranges between 50 Rs/month (Kannampalayam), 80 Rs/month (Nelamangala), 100 Rs/month (Palladam) and 120 Rs/month (Ramanagara), which works out to Rs. 3–10/m^3 across all towns[13]. This is comparable to the price of water in other towns in India at that level of consumption (CSE, 2012).

The hidden cost for those who are dependent on public standpipes, however, is the effort involved in waiting at the standpipe and carrying pots home. One indirect measure of the difficulty faced in obtaining water from standpipes is the frequency of the standpipes. Here, the towns in Tamil Nadu had a standpipe for every four dwellings in low-income areas, whereas the frequency in Nelamangala and Ramanagara varied. Some low-income areas in Nelamangala did have standpipes at every other doorstep, but both supply and standpipe frequency in Ramanagara was much more erratic, as per our observations.

To summarise, it appears that average supply during most of the year was adequate (if one uses the lower 70 lpcd norm) in three of the four towns, but is quite inadequate in Nelamangala. And the provision of free water at the standpipes and the low (and fixed) price of water for in-house connections has kept the water affordable. The systematic dual-sourcing in the Tamil Nadu towns is able to ensure that some better quality (river) water is available to all households for drinking and cooking purposes, but this approach is not used in Ramanagara and not feasible in Nelamangala because of lack of access to river water.

[12] The ULB in Nelamangala also supplied tanker water in summer months, but that was clearly not enough.

[13] Using an average household size of 5 and water use figures of 80–100 lpcd, because households with in-house connections consume more water.

Equity

Equity was assessed at three levels: between towns, between wards, and between households. Between towns, as Figure 3 shows, the inequity is clear: Nelamangala citizens get much less supply as compared to the other three towns. This is attributable to the unequal allocation of water that is imported from neighbouring basins. Whereas both towns in the Noyyal basin get significant (about 50%) of their supplies from imported river water[14], in the Arkavathy basin, only Ramanagara has access to such water, in spite of the fact that Nelamangala is less that 16 km away from the edge of Bengaluru metropolis, which imports 1,350 million litres per day (MLD) of river water[15]. The state of Tamil Nadu has not only constructed a number of pipelines to import river water from the Bhavani, Siruvani and Amaravathy into the Noyyal basin, but has also systematically ensured that all small towns (as well as villages en route of the pipeline) are supplied some river water from one or the other pipeline. In other words, Tamil Nadu state seems to have put in more efforts to distribute imported river water evenly across towns as compared to Karnataka state.

Our attempt to quantitatively estimate inter-ward variation within towns gave unreliable results. This is because in Nelamangala and Ramanagara towns, there was no easy mapping between OHT and ward, and there was significant inter-ward transfer in Kannampalayam and Palladam towns. What we have, therefore, are qualitative impressions based on interviews with residents, councillors and ULB staff. It was acknowledged that there were inter-ward variations. Some of these were attributed to (a) the capacities of overhead tanks not matching current population densities of the areas they service, (b) random variations in borewell yields (this variation is particularly high for deep borewells in hard rock aquifers), (c) variation in the presence of commercial users and their grabbing water meant for domestic consumers, and (d) newly added wards not having the connectivity with the OHTs in which where river water is stored. Some councillors argued that there was also systematic neglect of wards whose residents belong to socio-economically marginalised groups, but it was difficult, given our data, to tease apart this possibility from the effects of the above-mentioned 'physical' factors.

For understanding inter-household variation, we have systematic data only from the two Karnataka towns. These are presented in Table 4, and show that households dependent on standpipes alone may be at 40 lpcd whereas those with in-house connections and sumps may be at 47–60 lpcd in Nelamangala but above 75 lpcd in Ramanagara. Although the monitoring data comes from very small samples and therefore with large uncertainties, they match the results from more rigorous studies elsewhere (Kumpel *et al.*, 2017) which show a strong correlation between the extent of in-house storage and lpcd[16]. In other words, under intermittent supply, the wealth of a household broadly influences its ownership of storage assets such as sumps and this in turn significantly influences the ability to consume water. But it also appears that the presence of a sump makes a big difference to consumption only when there is relatively higher water supply, as in Ramanagara, as against when supply itself is poor, as in Nelamangala. So the major driver of inter-household variation is the ability to afford an in-house connection and a sump (the socio-economic factor at the bottom of Figure 1), perhaps compounded by the supply situation.

---

[14] 'Import' here referring to cross-basin transfers.

[15] And this pattern is true for the larger population of small towns in both basins.

[16] Note that Kumpel *et al.* measured all in-house storage, whereas we only noted presence/absence of a sump (not barrels or buckets), so ours is a much coarser estimate of in-house storage.

Table 4. Variation in connectivity, sump ownership and consumption.

Variable	Nelamangala	Ramanagara
% of households that are fully standpipe dependent (and their estimated consumption – pot survey)	14% (~40 lpcd)	10%
% of households with in-house connections (and their estimated consumption – monitoring)	86%	90% (~53 lpcd)
% of households with in-house connections and sumps (and their estimated consumption – monitoring)	77% (47–60 lpcd)	45% (75–120 lpcd)

Note: Connectivity and sump ownership data are from the large-sample source mix surveys (N = 300 and N = 600 for N and R respectively) and consumption estimates are from household-level monitoring (N = 5 and N = 8 respectively) for households with in-house connections, and 'pots collected' survey (N = 41) for fully standpipe dependent households.

Biophysical sustainability

Biophysical sustainability can be examined in terms of the trends in GW levels and the status of SW harvesting structures. All towns depend on GW to some extent. The sustainability of this resource is therefore crucial to sustaining water supply in these towns. The two basins in which the towns are located are undoubtedly facing groundwater depletion due to agricultural pumping (Lele *et al.*, 2013; Srinivasan *et al.*, 2014). It is therefore not surprising that the linesmen in all the towns reported that groundwater levels are dropping.

But the groundwater situation is not identical across all towns (see Table 5). The towns facing most groundwater sustainability issues are Nelamangala and Palladam. In terms of actual failure of borewells, Palladam had 13 borewells (8% of the total) failing in 2014–15. Although Nelamangala ULB staff could not give the exact number of borewell failures, they dug 12 new borewells that year, of which four did not yield any water. Typical borewell depths there were 750 ft (225 m), and the taluka (sub-district) itself has been notified as over-exploited by the Central GW Authority[17]. On the other hand, Kannampalayam and Ramanagara did not report any borewell failures as such, although linemen there did report declining groundwater levels.

This variation in groundwater depletion can be explained by a combination of geography and extraction patterns. Nelamangala is the heaviest user of groundwater and it is located in the upper reaches of the Arkavathy basin, which has seen severe groundwater depletion due to agricultural use and the consequent drying of the river (Srinivasan *et al.*, 2015). Ramanagara uses limited amounts of groundwater,

Table 5. Borewell failures in 2014–15.

Town	Nelamangala	Ramanagara	Kannampalayam	Palladam
Number of municipal borewells that failed in 2014–15 (% of total BWs in town)	12 replacement borewells dug (4 unsuccessful); total borewells 70	0 (0%)	0 (0%)	13 (8%)

Note: Nelamangala ULB staff could not give data on how many borewells had failed.

[17] http://www.cgwb.gov.in/CGWA/documents/CGWA%20Notifiaction%20dated%2027-11-2012.pdf.

and is located in the middle reach of the Arkavathy, where the river flows intermittently. In the Noyyal basin, Kannampalayam is next to the Noyyal River and immediately downstream of Coimbatore city, which releases more than 300 million litres per day of sewage into the river (Srinivasan *et al.*, 2014). Not surprisingly, the water table in Kannampalayam is shallow, with a number of functioning open wells. Palladam is, however ~11 km from the river, almost at the edge of the surface watershed, and (as seen in Figure 3) extracts much more groundwater than Kannampalayam.

The biophysical sustainability of imported river water is not an issue as such. SW is a flow resource, and as long as the rivers from where the water is imported flow, the water will be available to the towns through the bulk supply system. But the sustainability of local SW harvesting structures is relevant. Here, we see that SW harvesting and storage structures, i.e., erstwhile irrigation tanks that each town is endowed with, are in complete disrepair. Kannampalayam is not using its tank to harvest local rain-water runoff, but to collect effluents released by Coimbatore city for groundwater recharge. And the tanks in Nelamangala and Palladam towns are completely filled with sewage, rubble or solid waste (see, for example, Figure 4), whereas Ramanagara's tank is dry. This is particularly paradoxical, given that the import of river water is from a distance of over ~100 km.

Financial sustainability

The minimum criterion for financial sustainability is that O&M costs are covered by the water charges (Jayaramu *et al.*, 2015). Accordingly, we compared O&M costs with income from water charges for all the towns. The details are given in Table 6. It is clear that, in spite of uncertainties, Palladam town is breaking even but the other three ULBs are not even close to breaking even on O&M costs – their income from water charges ranges from 11% to 28% of these costs. And data from previous years (not presented here) show that this has consistently been the case.

Fig. 4. Status of Palladam tank.

Table 6. Annual expenditure and income related to water supply (in 10^5 Rupees unless mentioned otherwise).

Town		Nelamangala	Ramanagara	Kannampalayam	Palladam
Expenditure	River water charges	0	(Included below)	7.4	28.8
	Electricity	53.5	129.1	19.4	30.8
	Maintenance	68.0	89.5	72.8	19.3
	Salaries of staff	*Not available separately*	*Partly included in maintenance*	00.4	18.4
	Total	**121.5**	**218.6**	**100.0**	**97.3**
Income from water charges (all consumers)		**13.7**	**47.1**	**28.7**	**99.2**
Efficiency of water charges recovery		**~70%**	**~20%**	**97%**	**90%**
O&M cost of water supply Rs/m^3		33	19	15	6.9
Income per m^3		2.7	1.8	2.2	7
Budgetary deficit		**89%**	**78%**	**71%**	**nil**
Year of data		2012–13	2014–15	2012–13	2012–13

Notes: 1. Numbers pertain to the most recent year for which data were provided from the ULB/para-statal.
2. Income figures for Ramanagara are estimated assuming a 20% recovery rate, as informed by the engineer.
3. Expenditure on salaries of ULB staff who assist the para-statal in water distribution in Ramanagara were not available. Nor were electricity costs of pumping borewell water included. So actual costs in Ramanagara are significantly higher than estimates given here.
4. Salary costs in Kannampalayam are low because many contract staff salaries are included under maintenance.
5. 1 Indian Rupee ≈ 0.015 US$ during the study period.

There are multiple reasons for this outcome. First and foremost, water charge recovery is very poor in the Karnataka towns, ranging from 20% (Ramanagara) to 70% (Nelamangala), whereas it was above 90% in the Tamil Nadu towns. Second, the cost of water is higher for towns dependent on deep groundwater: in Nelamangala because it is entirely dependent upon pumping from deep (>600 ft) and low-yielding borewells, the cost is Rs 33/m^3 because of high electricity and pump maintenance costs. Whereas the cost of river water supply ranges from Rs 9/m^3 in the case of the Noyyal basin to Rs 15/m^3 in Ramanagara[18]. Third, a detailed examination of the maintenance costs showed that Kannampalayam's maintenance costs are high not just because of the salaries they include but also an inexplicably high expenditure on pump maintenance, which officials were not able to explain (and which suggests a mismanagement of funds).

How do these towns meet this continuous deficit? They use development grants provided by the state government under fiscal resource sharing policies towards covering these deficits, thereby reducing their regular development spending. Indiscriminate spending of 'drought relief funds' on digging new borewells to appease citizens rather than addressing the problem of groundwater sustainability in the long-term is another example of short-sighted policies.

The case of Palladam, however, shows that it is not impossible to meet O&M costs from water charges. Palladam has the lowest O&M costs (6.9 Rs/m^3) and the highest income from water charges (7 Rs/m^3). Discussions with their staff and comparison across towns suggest that in addition to a

[18] TWAD Board charges only Rs 4.5/m^3 to the small towns in Tamil Nadu, but their officials informed us that their actual O&M cost for bulk water supply is around 9 Rs/m^3. So when comparing cost of water across towns, we have taken the full cost of river water as 9 Rs/m^3, but when examining income-expenditure, we have used the lower billed amount.

tight control on maintenance costs, and a strong focus on water charge recovery, they have the highest tariff for in-house connections (Rs. 100/month as against Rs. 50–60/month in other towns) and similarly higher tariffs for commercial and industrial consumers. The last measure is certainly a crucial component of financial sustainability.

Responsiveness and accountability

We assessed responsiveness and accountability on four parameters, and summary 'grades' for each town are given in Table 7. The basis for these grades is as follows.

In terms of how the water supplying agency responded to day-to-day citizen complaints, the ULBs in Tamil Nadu had complaint books, which were actually in use, and interviews with citizens revealed that they had a satisfactory experience in terms of grievance redressal. Citizens had phone numbers of linemen, ULB staff and elected officials, which they did not hesitate to use, and which brought quick redressal. In the Karnataka towns, citizens in Nelamangala said that they call the councillor in case of distress and s/he communicates with the ULB staff to resolve the issue. In Ramanagara, however, citizens found the para-statal staff not very responsive, and councillors were unable to redress grievances quickly because the distribution was handled by the state-level para-statal, in which ultimate accountability was to their own head office in Bengaluru. At the same time, in Nelamangala, citizens reported having to pay some 'additional amounts' (tips or bribes) to the linemen to get water released during times of scarcity – clearly frontline staff taking unfair advantage of the high level of water scarcity in that town. Whereas in Ramanagara, the frontline staff were colluding with customers to provide additional 'illegal' connections (for which no monthly charges were being paid).

In terms of responding to summer-time scarcity, we found that both the ULBs in Tamil Nadu were more responsive and organized: they deployed their own tankers and moved water from functioning borewells to supply the service area of failed borewells. On the other hand, in Nelamangala, while the ULB did provide water through tankers during the summer, they did this in fairly irresponsible ways. Councillors blocked the executive officer's attempts to purchase a tanker for the ULB, and also blocked attempts to share water across wards; they preferred to spend scarce ULB funds on purchasing water from private tankers for augmenting the supply – reportedly due to their having a vested interest in those private tanker companies. They also spent special state funds made available for 'drought relief' on digging new borewells, rather than on measures to address groundwater sustainability in the long run. One councillor, however, used personal funds to dig a borewell and lay 'alternate' line of standpipes in his constituency. In Ramanagara, the para-statal passed on the responsibility of tanker supply during distress months to the ULB, which agreed to take on these costs because of the political sensitivity of the water issue.

Table 7. Broad assessment of responsiveness and accountability.

Indicator	Nelamangala	Ramanagara	Kannampalayam	Palladam
Response to day-to-day grievances	Fair	Poor	Good	Good
Response to summer shortages	Fair, but short-sighted	Poor, erratic	Good	Good
Record-keeping quality	Fair	Poor	Good	Good
Councillor involvement in long-term planning	Limited	None	Limited	Limited

As a basic form of accountability, we found that the quality of record-keeping in the four towns varied significantly. The two towns in Tamil Nadu maintained and provided detailed records on bulk water purchase, monthly electricity bills, water connections and charges collected, borewell pump details and electricity meters, etc. This also reflected in high levels of water charges recovery. Whereas both towns in Karnataka did a poor job of record-keeping – they did not have lists of households with in-house connections, electricity bills for municipal borewells could not be traced or did not match meter numbers at the borewell, borewell-wise pump details were not available, nor were details of water being pumped to each OHT. And this poor record-keeping reflected in poor water charges recovery. Even between the two towns in Karnataka, the record-keeping was surprisingly poorer in the para-statal-managed town (Ramanagara) than that in the ULB-managed system (Nelamangala) – the para-statal provided meagre data, and kept changing their estimate every few months.

In mid- to long-term planning, such as whether or not to give new connections, proposing new construction of OHTs and seeking funds from the state government for various projects, one sees the involvement of executive staff in the towns which have ULB-managed distribution. Elected officials are hardly involved, except perhaps the chairpersons of the ULBs. In the case of Ramanagara, where the water supply and distribution are managed by the para-statal, the planning process was completely opaque and engineer-controlled, and there was no mechanism for input from elected officials at all. The elected officials were, however, involved in decisions about setting water charges and about where to dig borewells when state funds became available.

Discussion

In the previous section, we have identified the broad patterns in the outcome variables and the intermediate decisions (proximate causes) that might have resulted in these variations. Here, we discuss how these might relate to the deeper causes, including the service modality in its institutional and technical dimensions, and the other factors in our framework.

First, overall water scarcity is explained by a combination of the biophysical endowment and the service modality. The biophysical endowment and socio-economic context of the towns clearly sets the baseline for this scarcity. All towns are growing rapidly, but are located in regions limited and declining surface and groundwater availability[19]. Groundwater alone cannot suffice, and is not the preferred source because of quality concerns. SW is essential, and access to it enables towns to at least come close to societal norms of average supply. At the same time, the socio-technical mode of managing these two sources is important. By embracing a dual-source policy and supplying river water and groundwater in parallel, the Tamil Nadu towns are able to ensure that higher quality water goes for higher quality end-uses and therefore spread the higher quality water more evenly. Households adapt to this systematic dual-sourcing, storing SW and GW separately and using them for different activities, thereby optimising the resource vis-à-vis end-uses. Simply augmenting overall supply by installing GW-based supply in some locations, as is the case in Ramanagara, does not help.

Is there a link between the systematic adoption of dual-sourcing and the institutional arrangement? We believe there is: the separation of roles between TWAD board (bulk supply of river water) and

[19] Kannampalayam's shallow groundwater is the result of sewage discharges from upstream Coimbatore city.

the ULB (distribution) also means that a uniquely engineering mindset that understands SW as the sole and proper water to be distributed, does not dominate at the distribution end. The objective remains supplying water, not just river water. We have observed this phenomenon in the metropolis of Bengaluru as well – the para-statal agency, being composed exclusively of engineers, only believes in supplying river water, and refuses to integrate groundwater into its paradigm of water supply. Even when it inherits public borewells as a result of an expansion in its jurisdiction to erstwhile small towns, it sees them as temporary arrangements until river water supply pipelines are put in place. Not surprisingly, private well ownership in cities such as Bengaluru is around 30% (Malghan *et al.*, 2016) but has remained below 5% in the small towns we studied in Tamal Nardu[20].

Second, inequality across households in water consumption is of course primarily a result of pre-existing economic class differences – a households' ability to afford the initial cost of an in-house connection and, in the context of intermittent supply, to store water for several days is crucial in getting access to more water. The question is how the governance system responds to these pre-existing differences. All towns have responded to socio-economic inequality through the concept of free supply at public standpipes, and increasing the density of such standpipes in the poorer neighbourhoods – but the town with the para-statal has done the poorest job here. And its separation of surface and groundwater supply regions has meant that poorer households also get only water of greater hardness.

Third, a fairly clear relationship does seem to emerge between the institutional arrangement and responsiveness and accountability in service delivery. Where ULBs do the distribution, there is greater responsiveness to day-to-day grievances, and also some local involvement or at least transparency about longer-term decision-making. This is not surprising, given that para-statal is only accountable to the state government, not the ULBs. But some of the differences between the three ULB-managed towns, however, appear to be related to the differences in the political and bureaucratic cultures of the two states: the better quality of record-keeping in the Tamil Nadu towns[21] for one, and the more competitive but less conflictual functioning of the councillors for another.

Fourth, the inter-town inequities cannot be explained by the factors in the framework: they seem to be driven by higher-level processes regarding allocation of water across towns and sectors, which are different in the two states. In Tamil Nadu, there appears to be an 'empowered committee' at the state level that takes decisions regarding all bulk supply projects[22], and there seems to have been a conscious decision that bulk supply must be shared across all towns and also with villages en route from the source to the town. In Karnataka, there seems to be no institutionalised mechanism. A Karnataka State Urban Water Council was mooted under a World Bank-aided project (Sustainable Development Department, 2011), but never created (K. P. Krishnan, personal communication).

Fifth, again, neither differences in service modalities nor in the political or administrative culture appear to affect the resource sustainability outcomes: groundwater depletion is ubiquitous, as is neglect of local water bodies. Admittedly, the groundwater depletion in these towns is driven not just by their own extraction, but also by over-extraction by surrounding farmers over decades. Pollution of local water bodies may be

---

[20] Unfortunately, it is not clear that the ULBs realize the benefits of dual-sourcing or of taking an integrated perspective. Kannampalayam and Palladam towns recently decided to not give any more in-house connections (because of limited supply of river water), but this will only drive new household into private borewell drilling, and easily lead to a Bengaluru-like situation where groundwater regulation then becomes impossible.

[21] Also corroborated by other small towns studied in the larger research project.

[22] http://www.twadboard.gov.in/twad/administrative_sanction.aspx.

caused by the inability of small towns to invest in sewerage, but given the neglect of sewerage infrastructure even in large cities such as Bengaluru (Jamwal *et al.*, 2014), the problem may be of mindset than just the lack of funds or technology. Interestingly, Kannampalayam town has tried to use the effluents from Coimbatore city to recharge its groundwater table by diverting and storing some of those flows in their SW tank (originally an irrigation reservoir). While the public health consequences of using untreated/poorly treated city effluents for groundwater recharge are likely to be negative, the effort indicates some awareness of the groundwater situation and a willingness to attempt recharge in some form (and perhaps a level of desperation).

Similarly, different modalities surprisingly do not seem to affect financial sustainability. One would have expected a para-statal to be 'stricter' in recovering its O&M costs, because it is not beholden to local electoral politics. But either by systematically under-charging for river water (as the TWAD board does) or neglecting to recover full O&M costs (as KUWSDB does), the para-statals seem to be driven by the perceived importance of local water supply in state-level electoral politics, as does the continued subsidy to towns to cover O&M costs from development funds. And of course ULB's are reluctant to raise water charges, because the councillors see their electoral fortunes dependent on low rates. But as Palladam's case shows, it is not politically impossible to raise water charges and ensure water charge recovery to a level that meets at least the O&M costs. It also suggests that metering may not be crucial to achieving financial sustainability, if basic costs and recoveries are managed well, a conclusion in line with the findings of Kumpel *et al.* (2017).

Conclusions

We began this study by pointing out that the research on small town water governance in India was very sparse, but that, unlike, say, small towns in many African countries, those in India are by and large characterised by nearly full coverage by government agencies and are above the absolute minimum of 25 lpcd. Consequently, the debate is not about how to interface public, private and community efforts, but about how to make these government agencies (or their combinations) more effective. And since the minimum threshold has been crossed, the concerns in policy-making should go beyond sheer adequacy and access to questions of equity, physical and financial sustainability, and responsiveness. By taking such a multi-dimensional approach to assessing outcomes, and by defining service modality in broad terms, we have managed to provide a more holistic picture of small town water governance in southern India. Our study perhaps provides the first (even if approximate) estimates of the contribution of groundwater to public supply, and also possibly a first-time comparison of multi-dimension outcomes across different institutional and socio-technical arrangements. The small town universe is vast and heterogeneous, even just in India, and so generalisations are risky. What we offer through this exploratory study that used an inductive approach are not exact policy prescriptions but rather hypotheses that require further testing.

An easy way to encapsulate the policy-relevant hypotheses that emerge from our analysis, across the two components of 'service modality', is the idea of 'match–don't-mix'. That is to say, a separation and matching of roles and responsibilities as well as of resources, rather than muddling or mixing them, is desirable. First, regarding institutional arrangements, and following Mugabi & Njiru (2006) in thinking at multiple levels:

- In terms of roles of different agencies, it appears that bulk supply of SW is best done by an engineering-oriented provincial para-statal agency or government department, whereas local groundwater supply and distribution of all water is best handled by a wing of the local elected government. The latter can provide more equitable and responsive water service delivery, whereas the former is set up for deployment of regional water resources (a point also made by Krishnan, 2006)[23]. This is not to say that ULB-led water management is a panacea – being answerable to users makes ULBs reluctant to raise water tariffs even to levels that would cover O&M costs, and rent-seeking by elected officials and their patronage politics (Nadhamuni, 2012) may be worse than the non-responsiveness of para-statal engineers. But closing the fiscal loop on ULBs (see below) rather than usurping their functions would be the way forward. This would also require ULBs to build their technical capabilities and staffing, something that has been recommended repeatedly in India (Rai & Singh, 2008; Perret *et al.*, 2014).

- Inter-town allocation of SW has crucial implications for meeting adequacy norms in individual small towns, and therefore requires much more 'due process' at the state level to ensure fair allocation. The debate is not whether imports of SW is essential or not (as Mukherjee *et al.*, 2010 seem to suggest), but about how to fairly allocate what will be an increasingly scarce, expensive and contested resource.
- The state (provincial) governments have an important role to play, but they need to keep away from meddling in lower-level functions such as staff deployment or pushing para-statals into water distribution and engage more systematically with higher-level functions. These functions include not only allocating river water fairly and transparently across towns and villages, but also replacing state financial grants with more devolution of taxation powers to make ULBs more self-financially sufficient, and building ULB capacities to address questions of groundwater sustainability and wastewater treatment and reuse.

Second, with respect to the socio-technical dimension of service modality:

- Consciously adopting a dual-source (groundwater plus SW) approach allows the matching of end-use needs to quality of water, and provides more equitable access to higher quality water, in addition to augmenting overall availability. Ironically, although this approach is being practised in many of Tamil Nadu's small towns, it is not clear whether it has in fact been institutionalised[24]. The idea that 'GW is an inferior resource, to be used as a last resort' or only until river water pipelines are completed originates in a surface water-oriented water engineering community. But in fact, GW provides an important supplement and buffer and also acts as a barometer of the state of local resources[25]. Dual-sourcing (as a part of integrated water management) needs to be institutionalised, and it is more likely to happen within ULBs than within engineering-oriented para-statals. This may involve more investments (dual piping) and complexity in operations, but will make supply more sustainable and use more efficient (by matching sources to end-uses).
- At the same time, efforts will be required to regenerate and augment local resources, through rooftop and lake-based rainwater harvesting, and wastewater treatment and reuse. But this will require shifting away from the illusion that continuous expansion of cross-basin or long-distance SW transfers is possible, and similarly continuous expenditure of public funds on simply drilling new borewells or buying private tanker water for summer supply is possible.

---

[23] Indeed, there is a growing feeling that even large cities should bring para-statals under municipal control (Ravindra, 2010).

[24] The latest projects initiated in Coimbatore city involve increasing imports of river water to meet needs in newly added areas, rather than spreading existing imported water equitably across the whole town (Durba Biswas, pers. comm.).

[25] 'Wells talk to people' (S. Vishwanath, quoted in Sitlhou, 2013).

In a rapidly urbanizing countries such as India, with villages turning into small towns and existing small towns growing quickly, local urban governments are always playing catch up, struggling to provide just basic service delivery. Provincial governments are tempted to deploy engineering para-statals to quickly 'solve' problems, often citing the lack of technical capability or under-staffing in the ULBs (De Bercegol & Gowda, 2013; Atkins *et al.*, 2014). Towns are caught between the reality of GW dependence and their aspirations for complete SW supply that large metropolises seemingly have achieved, aspirations promoted by engineers who only understand SW supply. But these towns have the advantage of being earlier on the curve, and need not repeat the mistakes made by large metropolises. Expanding the goals of both policy and research to think in terms of multi-dimensional water governance rather than delivery of an average quantum of water, and bringing a socio-technical-cum-institutional idea of service modality, along with other factors, to the research could support this process towards more sustainable and equitable development.

Acknowledgements

We are grateful to the citizens, municipal staff and elected officials of the towns of Nelamangala, Ramanagara, Kannampalayam and Palladam, and officials of TWAD Board and KUWSDB for their cooperation and support. We also thank the Directorates of Municipal Administration and Town Panchayats in the two states and Professors Suresh Kumar and S. Paramasivam from Tamil Nadu Agricultural University for facilitating this research. GIS support was provided by the ATREE Eco-informatics laboratory and Parvathy Menon and Rahul Varier from ATREE helped in field work. Comments received from the guest editors and other contributors in this Special Issue and two anonymous referees were very valuable in sharpening the arguments, and are gratefully acknowledged.

Financial support for this research came from International Development Research Centre (IDRC) grant no. 107086-001 under their Climate Change & Water programme.

References

Adank, M., Darteh, B., Moriarty, P., Osei-Tutu, H., Assan, D. & van Rooijen, D. (2011). *Towards Integrated Urban Water Management in the Greater Accra Metropolitan Area*. SWITCH / RCN Ghana, Accra.

Anand, N. (2011). PRESSURE: The poliTechnics of water supply in Mumbai. *Cultural Anthropology 26*(4), 542–564.

Atkins, University College London & Indian Institute of Human Settlements (2014). Future Proofing Indian Cities – Final Urban Diagnostic for Bangalore. CDKN Project Report. WS Atkins International Ltd, London.

Bakker, K., Kooy, M., Shofiani, N. E. & Martijn, E.-J. (2008). Governance failure: rethinking the institutional dimensions of urban water supply to poor households. *World Development 36*(10), 1891–1915.

Cain, A. & Mulenga, M. (2009). *Water Service Provision for the Peri-Urban Poor in Postconflict Angola, Human Settlements Working Paper Series No. Water-6*. International Institute for Environment and Development (IIED), London.

Caplan, K. & Harvey, E. (2010). *Small Town Water and Sanitation Delivery: Taking A Wider View*. WaterAid, London.

CSE (2012). *Excreta Matters: State of India's Environment : A Citizens' Report-7 (Vol. 1)*. Centre for Science & Environment, New Delhi.

De Bercegol, R. & Gowda, S. (2013). Decentralised Water Management in small towns of India. *Proceedings of the 13th Edition of World Wide Workshop for Young Environmental Scientists (WWW-YES-2013)-Urban Waters: Resource or Risks?* (3-7 June 2013 at Arcueil, France). D. Thevenot (ed.). HAL-ENPC, Arcueil, France. Available from: https://hal-enpc.archives-ouvertes.fr/hal-00843101.

Fragano, F. (2001). Management models for small towns: community water board in Itagua, Paraguay. In: *Case Studies on Decentralization of Water Supply and Sanitation Services in Latin America.* Fragano, F., Linares, C., Lockwood, H., Rivera, D., Trevett, A. & Yepes, G. (eds). EHP Strategic Paper, No. 1. US Agency for International Development, Washington, DC, pp. 29–43.

Gupta, D. (2012). Beyond the metropolis. *Seminar 629*, 73–76.

GWP TAC (2000). *Integrated Water Resources Management., TAC Background Paper No. 4.* Global Water Partnership Technical Advisory Committee, Stockholm.

Indian Institute for Human Settlements (2014). *IIHS – Future Proofing Indian Cities – Final Urban Diagnostic for Bangalore.* WS Atkins International Ltd, p. 45.

Jamwal, P., Thomas, B. K., Lele, S. & Srinivasan, V. (2014). Addressing water stress through wastewater reuse: Complexities and challenges in Bangalore, India. In: *Proceedings of the Resilient Cities 2014 Congress, 5th Global Forum on Urban Resilience & Adaptation.* Organized by ICLEI, Bonn, Germany. Available from http://resilient-cities.iclei.org/fileadmin/sites/resilient-cities/files/Resilient_Cities_2014/RC2014_Congress_Proceedings/RC2014_congress_proceedings_Jamwal.pdf.

Jayaramu, K. P., Kumar, B. M. & Rashmi, K. K. P. (2015). Improving cost recovery in urban water supply service – an experience from India. *Water Policy 18*, 685–707.

Joy, K. J., Paranjape, S., Kiran Kumar, A. K., Lele, R. & Adagale, R. (2004). *Watershed Development Review: Issues and Prospects, CISED Technical Report.* Centre for Interdisciplinary Studies in Environment and Development, Bangalore.

Krishnan, K. P. (2006). Urban local bodies and district administration: an empirical examination of some issues relating to urban water supply in Karnataka. In: *Local Governance in India: Decentralization and Beyond.* Gopal Jayal, N., Prakash, A. & Sharma, P. K. (eds). New Delhi: Oxford University Press, pp. 296–323.

Kumpel, E., Woelfle-Erskine, C., Ray, I. & Nelson, K. L. (2017). Measuring household consumption and waste in unmetered, intermittent piped water systems. *Water Resources Research 53*(1), 302–315.

Lele, S., Srinivasan, V., Jamwal, P., Thomas, B. K., Eswar, M. & Zuhail, T. M. (2013). *Water Management in Arkavathy Basin: A Situation Analysis.* Environment & Development Discussion Paper No. 1. Ashoka Trust for Research in Ecology and the Environment, Bangalore.

Malghan, D., Mehta, V. K., Goswami, R., Nehra, A., Lele, S., Kemp-Benedict, E., Muddu, S., Mathur, C., Srinivasan, V., Wong, D., Gautam, M., Vivek, V. & Sharma, D. (2016). Bengaluru's domestic water consumption: Preliminary results for adequacy and equity. In: Paper presented at the *National Dissemination Workshop on Adapting to Climate Change in Urbanising Watersheds*, New Delhi, organized by Ashoka Trust for Research in Ecology and the Environment in New Delhi on 22–23 August 2016.

Mathur, M. P., Chandra, R., Singh, S. & Chattopadhyay, B. (2007). *Norms and Standards of Municipal Basic Services in India.* Working Paper WP 07-01. National Institute of Urban Affairs, New Delhi. Working Paper WP 07-01.

McKenzie, D. & Ray, I. (2009). Urban water supply in India: status, reform options and possible lessons. *Water Policy 11*(4), 442–460.

Mugabi, J. & Njiru, C. (2006). Managing water services in small towns: challenges and reform issues for low-income countries. *Journal of Urban Planning and Development 132*(4), 187–192.

Mukherjee, S., Shah, Z. & Kumar, M. D. (2010). Sustaining urban water supplies in India: increasing role of large reservoirs. *Water Resources Management 24*(10), 2035–2055.

Mundle, S., Chowdhury, S. & Satadru, S. (2016). *Governance Performance of Indian States: 2001-02 and 2011-12,* NIPFP Working Paper No. 164. National Institute of Public Finance and Policy, New Delhi.

Nadhamuni, S. (2012). *An Approach to Integrated Urban Water Management (IUWM): The Mulbagal Experience.* Arghyam, Bangalore.

Pahl-Wostl, C., Lebel, L., Knieper, C. & Nikitina, E. (2012). From applying panaceas to mastering complexity: toward adaptive water governance in river basins. *Environmental Science & Policy 23*, 24–34.

Perret, L., Joshi, R., Kantak, A., Poddar, R., Biyani, A., Khandelwal, R. J. & Fayoux, R. (2014). *Panorama of the Urban and Municipal Sector in India,* Focales No. 20. Agence Francaise de Developpement, New Delhi.

Rai, N. & Singh, A. K. (2008). *Decentralized Urban Governance in India.* Regional Centre for Urban and Environmental Studies, Lucknow.

Raju, K. V., Praveen, N. & Anand, B. K. (2004). Groundwater in urban market: Can it sustain? A case study of Kolar city in south India. Presented at the Market Development of Water & Waste Technologies Through Environmental Economics. 28 May 2004 in Paris. MEED SA and CERNA, Paris. Available from: www.cerna.ensmp.fr/Documents/cerna_globalisation/Raju-paris.pdf.

Rautanen, S. L. & White, P. (2013). Using every drop – experiences of good local water governance and multiple-use water services for food security in far-western Nepal. *Aquatic Procedia*. At the Confluence – Selection from the 2012 World Water Week in Stockholm. *1*, 120–129.

Ravindra, A. (2010). Should elected civic body take over water supply concerns? Times of India. Bangalore. Available from: http://epaper.timesofindia.com/Default/Layout/Includes/TOINEW/ArtWin.asp?From=Archive&Skin=TOINEW&BaseHref=TOIBG%2F2010%2F06%2F02&ViewMode=HTML&EntityId=Ar00200&AppName=1

Rusca, M. & Schwartz, K. (2012). Divergent sources of legitimacy: a case study of international NGOs in the water services sector in Lilongwe and Maputo. *Journal of Southern African Studies 38*(3), 681–697.

Sangameswaran, P., Madhav, R. & D'Rozario, C. (2008). 24/7, 'Privatisation' and water reform: insights from Hubli-Dharwad. *Economic and Political Weekly 43*(14), 60–67.

Shah, M. (2013). Water: towards a paradigm shift in the Twelfth Plan. *Economic & Political Weekly 48*(3), 41–52.

Shah, M. (2014). *Urban Water Systems in India: A Way Forward*. ICRIER, New Delhi.

Sitlhou, M. (2013). Water matters, not just on World Water Day. *The Alternative*. Available from: http://www.thealternative.in/lifestyle/water-matters-not-just-on-world-water-day/.

Srinivasan, V. (2008). *An Integrated Framework for Analysis of Water Supply Strategies in A Developing City: Chennai, India*. Doctor of Philosophy Dissertation Thesis, Stanford University, Stanford, CA.

Srinivasan, V., Kumar, S. D., Chinnasamy, P., Sulagna, S. & Lele, S. (2014). *Water Management in the Noyyal River Basin: A Situation Analysis,* Environment and Development Discussion Paper 2. Ashoka Trust for Research in Ecology and the Environment, Bangalore.

Srinivasan, V., Thompson, S., Madhyastha, K., Penny, G., Jeremiah, K. & Lele, S. (2015). Why is the Arkavathy River drying? A multiple hypothesis approach in a data scarce region. *Hydrology and Earth System Sciences 19*, 1905–1917.

Subbaraman, R., Shitole, S., Shitole, T., Sawant, K., Obrien, J., Bloom, D. E. & Patil-Deshmukh, A. (2013). The social ecology of water in a Mumbai slum: failures in water quality, quantity, and reliability. *BMC Public Health 13*(1), 173–186.

Sustainable Development Department (2011). *Implementation Completion and Results Report (ibrd-47300) on A Loan in the Amount of Us$ 39.5 Million to the Government of India for the Karnataka Urban Water Sector Improvement Project, No. ICR00001950*. The World Bank, Washington, DC.

Tungpalan, R. G. (2009). Small towns water supply in the Philippines. In: Presentation at the WSP Small Towns Workshop, Vienna on 8 June 2009.

(de)Constructing the conditions for private sector involvement in small towns' water supply systems in Mozambique: policy implications

M. Tutusaus[a], P. Cardoso[b] and J. Vonk[c]

[a]*Corresponding author. IHE Delft Institute for Water Education, Westvest 7, Delft 2611AX, The Netherlands*
E-mail: m.tutusaus@un-ihe.org

[b]*Collins Sistemas de Água, Lda. Rua Kamba Simango n.398R/C Bairro da Polana Cimento, Maputo, Mozambique*

[c]*Vitens Evides International, Reactorweg 47, Utrecht 3542 AD, The Netherlands*

Abstract

Contracts have been developed to govern the relationship between principal and agent. These contracts are necessary for the principal to exert control over the agent assuming that the interest of both parties are at odds and information asymmetries exist. In the cases of private sector involvement in the expansion of water services in small towns in Mozambique we review in this paper we propose a revised approach to these assumptions. We argue that given the conditions present in small towns a different relationship is developed between private operators and asset holder than it would be the case in bigger, longer contracts for bigger urban centers. In this relationship much more information is exchanged and the objectives are shared, often making the conditions of the contract irrelevant for understanding how services have and will develop, giving more space to relational contracts and other informal arrangements.

Keywords: Contract; Small towns; Private sector involvement; Water services

1. Introduction

In small urban centers the existing coverage of basic public services is 'neither adequate nor far-reaching' (Brockerhoff and Brennan, 1998 in Cohen, 2006). The incidence of disparity in access to public services is further accentuated in Sub-Saharan Africa, where almost 50% of households in urban areas under 100,000 inhabitants lack access to piped water, waste disposal, electricity and schools (Cohen, 2006). These small urban centers in rural regions lack the infrastructural facilities or communication linkages, which prevent them from benefiting from regional growth (Rondinelli, 1983). Access to

doi: 10.2166/wp.2018.003

water, albeit only including services by formal providers, is lowest in those settlements with the lowest population (Ferre *et al.*, 2012).

In order to address these situations, the water community has brought forward different technological and, mostly, managerial solutions thought most suitable for rural realities. In the last two decades, community based management, through Community Based Organizations (CBOs), became widespread and the most common model for management of rural and village water supplies across sub-Saharan Africa (Moriarty *et al.*, 2002; Harvey & Reed, 2007, Lockwood & LeGouais, 2011). CBOs for water services have been promoted in many countries as they are argued to provide a structure that enhances ownership and empowers the recipient community. However, the successful implementation of these models also proved challenging and when successful only due to context-specific situations that hardly related to model but rather to local circumstances (Moriarty *et al.*, 2002; Harvey & Reed, 2007; Lockwood & Le Gouais, 2011). As an alternative to community-based management, forms of delegated management to (local) private operators through different Public-Private Partnerships (PPPs) arrangements have increasingly come into vogue (WSP, 2010).

1.1. The onset of small-scale private operators

Engagement of private sector such as the involvement of private operations in daily system management and operations, dates as back to the 1990s. In an attempt to restrict public borrowing and expenditure, donors and development banks promoted the private operators as a solution to replace the predominance of public organizations in the sector. In this period influential donors, such as the World Bank, highlighted the rampant 'pitiable' performance of the public sector and strongly advocated for the involvement of the private sector rather than concentrating efforts on other types of reforms aimed at strengthening the role of the public sector in the provision of services (World Bank, 1994). As part of this advocacy, private sector actors were praised for their capacity to act as new sources of non-public investments and their ability to improve access and quality of water supply (Prasad, 2006).

The combination of disappointing experiences of CBOs in sustaining services in rural and small village water supply systems and the advocacy for private sector involvement gave rise to the promotion of small private operators for the expansion of services in small towns. The (local) private sector quickly became a preferred alternative to community-based management. The local private sector was assumed to have the capacity to operate professionally and have the right incentives to ensure the desired level of services. This would, consequently, lead to efficiency improvements in the system, increasing coverage and ultimately enhancing service provision altogether (Harvey & Reed, 2006; Harvey & Reed, 2007; Adinyira *et al.*, 2010; Lockwood & Le Gouais, 2011; Ameayw Effah *et al.*, 2015). The promotion of small private operators as water operators in small towns mirrors the trend in urban settings, where the local private sector is presented as a viable alternative to increase access to those communities that bigger (public and private) operators have not been able to reach. These operators of small-scale systems are viewed as performing a bridging function to the communities which government has failed to reach through centralized water supply systems (Mehta *et al.*, 2007). As a result, these operators, once termed 'informal providers', are no longer seen as a nuisance, but rather as a long term partner with which to extend and improve water supply (Schaub-Jones, 2008). As part of this increasing recognition, these small private providers are seen as 'gap filers' and 'pioneers', providing services in those places where other (public) alternatives were not reachable. In some cases these providers have subsequently become sub-concessionaires of the formal utility (Kariuki *et al.*, 2006).

The promotion of small-scale private operators has increased in recent years due to disappointing progress made in rural water access. In many countries progress in rural water supply significantly lagged behind urban water coverage as eight of ten persons without adequate access to drinking water live in rural areas (WHO, 2015). Furthermore, service provision in small towns has moved beyond become more complex in recent years. Providing services through boreholes and hand pumps is not enough. Small towns require small water supply systems with on-premise access. These increased complexities of small town water provision make private operators more suitable than the 'traditional' community-led management practices to provide services (Larbi, 2005; Lockwood & Le Gouais, 2011). These complexities are associated with the sophistication of systems, that private operators are assumed to be better equipped to deal with, or the implementation of 'professional' billing collection systems and reporting. As a result 'the domestic private sector has emerged in many developing countries as a promising partner for addressing the challenges of water and sanitation provision in rural and small towns' (Ndaw, 2016: 5). National policies increasingly employ language that fosters the participation of private operators, rather than municipal services or CBOs. The option to partner with private operators is explicitly promoted in national strategies in Benin, Burkina Faso, Mali, Mauritania, Mozambique, Niger, Uganda and Rwanda (WSP, 2010). Official PPP statistics, however, fail to capture the involvement of the private sector in managing small piped water systems in Africa. Nevertheless, it is estimated that approximately one-third of small piped water systems in the countries mentioned above are under the management of private operators (World Bank, 2014: 7).

Despite the interest from donors and governments, documentation on the involvement of these operators is limited (Kleemeier, 2010; WSP, 2010; Ndaw, 2016). What is written so far about private sector involvement highlights potential challenges small towns pose to the viable introduction of private operators in these settlements (Hirn, 2013). A few broad assertions have been made, however. The first is that, in comparison with large cities, the customer base in these towns is believed to be insufficient to achieve economies of scale. Moreover, this is compounded by the distribution of the population within the town, which makes achieving economies of scale even more difficult (Annez, 2006; Pilgrim *et al.*, 2007). Furthermore, the economic development of these urban settlements, the welfare system, and the economic possibilities of the user itself severely limit the possibilities of a private operator to generate sufficient revenue (Hopkins, 2003; Moriarty *et al.*, 2002). As a result, challenges in small towns are normally summarized in three factors (Mara & Alabaster, 2008; Mugabi & Njiru, 2006; WSP, 2010; Moriarty *et al.*, 2013; Adank, 2013; Ndaw, 2016): (a) high cost of developing infrastructure and operating services to provide the dispersed population of small towns (lack of economies of density) in combination with a potential low revenue base; (b) limited technical capacity that is locally available within the town; and (c) the dynamic and diversity of customer segments due to the rapid population growth and other demographic changes. Given these factors, favorable conditions for private operators to establish themselves and operate fully functional installations require guarantees that these providers generate sufficient revenue. Tendering of partially functional or abandoned water systems will result in stranded tenders. Second, due to the demographic fluctuations in these towns, the operators in charge should demonstrate their capacity to address increasing technical complexities of the water system. Lastly, private operators should be allowed flexibility to bring in their own funds to expand the water supply system (and their customer base) as it is only through service expansion that their business can truly develop (Kleemeier & Lockwood, 2012).

1.2. Incomplete contracts and agent relationships

Underlying PPPs is a formal contract, which is signed in order to manage and govern the relationship between the buyer and seller of services. By stipulating promises or obligations to perform particular actions in the future (Poppo & Zenger, 2002), formal contracts essentially represent 'organizing mechanisms' (Vincent-Jones, 2000). With the involvement of the private sector and associated commercialization and professionalization of water services, formal contracts in the water services sector have become more prevalent. As such, formal contracts are becoming increasingly prominent as the organizing mechanism for the delivery of water services in small towns and formalizing the relationship between local private sector and government agencies.

Contracts that organize the provision of water come in a large variety of shapes and forms. The greater the complexity of an activity subject to be organized through a contract, the more detailed the contract necessarily becomes. Complex contracts 'may detail roles and responsibilities to be performed, specify procedures for monitoring and penalties for noncompliance, and, most importantly, determine outcomes or outputs to be delivered' (Poppo & Zenger, 2002: 708; see also Cooper, 2003). Water services in small towns are increasingly becoming more complex. These complexities are derived from the increasingly elaborate technical requirements of the systems, the requirements imposed on operators to formally report on performance, and the control exerted over these systems as regulators ensure more and more people have access to water. As a result the contracts organizing the provision of water services are also relatively complex contracts.

It is however impossible to develop contracts that will cover and foresee all contingencies that could possibly materialize. When contracts are unable to cover all the contingencies that may occur these contracts could be considered as 'incomplete' (Lane, 2000). Small towns include a diversity of demographic, socio-economic and bio-physical conditions. Developing contracts which covers such an array of conditions is virtually impossible. Given the complexity of water provisioning and the wide array of conditions to which the complex contract is applied, a blanket application of a model contract for water delivery in small towns is most likely going to be incomplete.

Complex contracts are subject to agency relationships. The principal-agent model is 'in essence a theory about contractual relationships between buyers and sellers'. In such a contract both parties clearly specify what the seller of the good/service (the agent) should do and what the buyer of the good/service (the principal) must do in return (Waterman & Meier, 1998: 174). The key problem in the agency relationships is to devise a contract that motivates the agent to work for the principal at the same time that the principal is prepared to pay a (fair) compensation corresponding to the efforts of the agent (Lane 2001: 31). The contract is required in this relationship as the two are assumed to have diverging interests and as 'rational utility maximizers' each and every one of them may engage in actions that may be pernicious to the other. This problem is exacerbated by the assumption that in such an exchange the agent, in our case the provider of services or water operator, enjoys a comparative advantage to the principal, the government. This advantage is derived from the access to information from the operations of the system that is not available to the principal. What results is an 'information asymmetry' in favor of the agent. Contracts are to address this problem as they allow the principal to 'manipulate and mold the behavior of agents so that they will act in a manner consistent with the principal's preferences' (Waterman & Meier, 1998: 174).

The contracts implemented between asset holder and operators in Mozambique, which we are analyzing in this paper, have raised certain issues derived from the complexities highlighted and leads to the question of which and how suitable contracts are at organizing mechanisms for the relationship. In this article we aim to address this question by studying the development of small-scale private operators in Mozambique. The main argument of this article is that contracts, under well understood conditions, can and should be a suitable way of organizing water services in small towns. An important reason for this is that the cases from Mozambique suggest it is possible to align the interests of the principal and the agent, effectively mitigating the principal-agent dilemma. At the same time, the contracts applied in Mozambique also suggest that these contracts require a degree of flexibility they currently do not have. Moreover, the contracts are incomplete and thus need to be complemented by more relationships of trust between the operators and the government.

2. Methodology

In analyzing this case we employ an analytical framework developed for the evaluation of concessions for the expansion of services to unserved areas in large urban settlements. We realize that small and large urban settlements could behave differently. Having said that, we believe that the analytical framework we employ would apply to contracts of all types and application. The elements of analysis are conceptualized in the performance requirements of the agent, as it is the performance requirements in a contractual arrangement that define what an agent may and must do (or not do) (Komives, 2001: 64; Delmon, 2012). We argue that the use of this framework allows for a better understanding and appreciation of the business strategy for the small-scale private operator. It is also through the application of this framework that we understand what the objectives the implementing agency are as they are the ones defining the requirements around which the framework revolves. In this framework we identify four key elements:

1. Type of and quality of service the agent may and must provide;
2. Number of connections the agent must make;
3. Timing of connections to be installed; and
4. Location and clients the agents may and must service.

These elements determine the cost structure of the water system as it follows from the required level of service level, including quantity, quality and continuity of service. Also it stipulates the potential market to cover which in its turn indicates the potential generation of revenue, taking into consideration the operator's possibilities of choosing to whom and where to expand. This is likely influenced by the affordability and willingness to pay for services of potential customers, as well as the reach of existing service delivery infrastructure. Similarly, when and how the connections need to be developed indicates the pace of service coverage extension. This also translates into costs that the operator would have to recover if they were obliged to expand infrastructure or capacity.

The data used in this paper were collected during fieldwork activities from June 2015 to February 2016. The data have been collected through 29 semi-structured interviews with relevant stakeholders. Interviewed respondents included representatives of the implementing agency (AIAS) in the sector (23), representatives of Directorate of Water (DNA) (3), the regulator CRA (4), technical advisors (5) and small-scale private operators (5). The data analyzed are derived from nine different systems. For three of them (Moamba, Caia and Manjacaze), local authorities such as the Town Council or the regional representation of the Ministry of Public Works and Water Resources were also interviewed to understand their position concerning the involvement of the private sector in the provision of services. The operator in Moamba is currently operating in six other towns. Moamba was the first system to be tendered in 2013, it has 5,000[1] inhabitants and in 2013 approximately 50% of the population was connected to the system through household connections or in-yard connections. The other systems managed by the same operator vary in size between 8,000 (Homoine) to 43,000 (Mopeia). The operator in Caia is only operating one system under AIAS. The contract for Caia was tendered and at the moment of signature the population was estimated to be about 25,000 inhabitants and 15% had access to water. Manjacaze was only signed in 2016. At that time, this town had about 10,000 inhabitations and only 8% of the population had access to water. This last operator manages two more systems: one in Nametil (65,000 inhabitants and 11% coverage), and one in Espungabera (6,000 inhabitants and 79% coverage).

Official documents such as contracts and regulations, and consultancy reports have been used to complement the data collection process. Additionally, data such as performance in relation to contractual KPIs have been accessed through the involvement of two of the authors in the actual implementation of contracts with private operators in small towns in Mozambique.

3. Water services in small towns in Mozambique

In Mozambique, the Ministry of Public Works, Housing and Water Resources (MPOHRH) through the National Water and Sanitation Directorate (DNAAS) is responsible for strategic planning and management of the sector (Figure 1). In case a municipal council exists, it is this body that remains responsible for guaranteeing water services. The municipality may decide to delegate these services to a third party or deliver those using municipal staff or creating a separate municipal company. In the absence of competent municipal services, the government is represented in the provinces by the Provincial Offices of the Ministry (DPOHRH) and at district level by the offices of Infrastructural

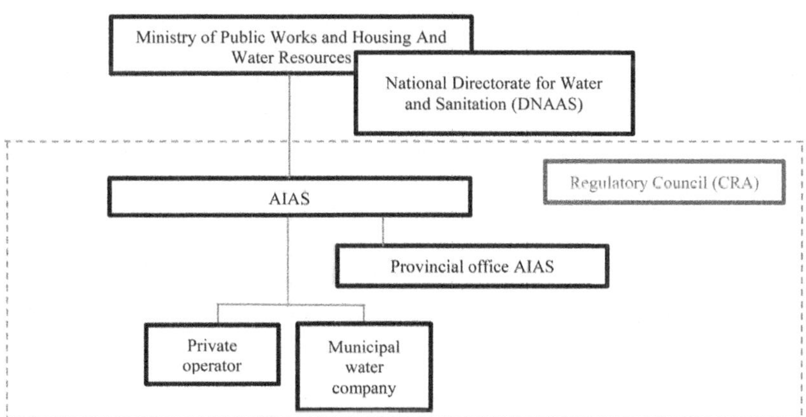

Fig 1. *Source:* Adjusted from WSP-AIAS (2015).

---

[1] All population figures are projected from the last census carried out in 2007.

Development (SDPI). These last offices support the development of water services in remote areas of the country.

In 1998, the government created the Delegated Management Framework (Decree 72/98). The DMF was to give an impulse to decentralization efforts in the water sector by allowing the possibility to grant management, lease or concession contracts to private companies, as a temporary measure to retrieve services from the public domain. Under this reform a new organization (FIPAG) was founded to act as asset holder and fund management for the development of water infrastructure. FIPAG was to enter in contracts with private operators for the development of water services in the larger urban areas of the country such as Maputo, Beira, Pemba or Nampula. During this same reform, an independent regulator CRA (Conselho de Regulação de Água) was set up to supervise the performance of those private operators engaged in contracts with FIPAG. After the delegation of services in large cities was consolidated, the Government of Mozambique extended the DMF to small towns creating the Administration for Water Supply and Sanitation Infrastructure (AIAS). Up until that moment, small towns that had not fallen under the mandate of FIPAG, but received the assistance of the National Directorate of Water (DNA) for the development of water works and other technical support through the provincial offices of the Ministry of Public Works. Systems remained solely managed by municipalities as stipulated by the National Water Law. The mandate of AIAS was designed to promote autonomous, efficient and financially viable management of water systems through the delegation of operations to the private sector, or other entities, to initially around 130 small towns[2] (Decree n. 19/2009 13 May 2009). The selection of small towns was rather ambiguous as no clear criteria were used for differentiating AIAS towns from other towns. As a result, the resulting list of AIAS towns is somewhat heterogeneous in nature (see Schwartz *et al.*, 2017). The towns assigned to AIAS range in population size from 2,430 (Nova Mambone–Inhambane) to 207.543 (Mocuba–Zambezia) inhabitants. The average population of AIAS towns is 25,000 inhabitants. AIAS manages the investment plans for the rehabilitation and construction of existing and new water systems, launches public tenders to encourage the private sector to participate in the operations of these systems and supervises the implementation of the contract. The selection of towns assigned to AIAS is not only diverse in size, but there is also a great divergence in both the state of the infrastructure as well as the extent to which water is already provided in these towns. In 2013, the average coverage in the towns assigned could vary from 6.1% in the province of Niassa or Sofala, to 38.1% in the province of Gaza. However, these divergences are also registered within the same province. For example, Gaza towns, such as Massingir or Eduardo Mondlane, reported access rates of 67.9% and 83.5% respectively, whilst the town of Manjacaze reported an access rate of 8.4%. Having said that, the general state of access to water in the selected towns for AIAS can be considered poor and only in very few exceptions (13 out of 130) exceeds 50%. The regulator CRA supervises and approves the implementation of water tariffs as well as the performance of operators. In the case of small towns, the monitoring responsibilities is delegated and is effectively carried out by local committees formed by local representation of users and local authorities. However, such committees had only been formed in a handful of towns by early 2016.

AIAS is currently engaged in 24 contracts with private operators. Each contract is linked to one single system. However, some operators have accumulated the management of more than one system. AIAS works in partnership with 14 different operators. One operator currently manages eight systems, and two

[2] Authors' translation.

other operators manage two or three systems. The other 11 are operators of one single system. On paper, all contracts are designed the same with the same requirements and conditions. In practice, some contracts are more specific than others, stipulating to a greater extent the expected coverage expansion and other KPIs such as service hours or billing and collection. The contracts are signed between the operator and the Directorate of AIAS. The subsequent monitoring is done via AIAS with local accountability committees formed by CRA at the local level, and the general oversight CRA exerts over the exercise of AIAS.

4. AIAS and the local private sector in partnership

Since its inception, AIAS has envisioned the expansion of services in small towns through the rapid engagement of private operators. Their standard procedure before an operator actually starts managing and operating a system begins with the rehabilitation or construction of the water system in the selected town. AIAS allocates a budget for these developments from its own funds channeled from the MPOHRH and through funds received from donors. Once the construction and rehabilitation work is set in motion, AIAS issues a tender call for private operators to bid on specific systems. For each tender a contract is granted to a private operator for an initial period of five years with an option of renewal for another two. The total maximum length of a contract is thus seven years. Once the contract is granted the operator is involved in the final stages of construction and/or rehabilitation works. This allows the operator to gain insights into the system that will become useful in the operation of the system. This was initially not the case but AIAS modified this part of the process due to a series of problems encountered with the initial system startups. Even though technical and operational characteristics are almost always clearly specified, operational realities have been known to deviate leading performance levels to differ from initial expectations. As a result, some private operators had to rapidly adjust their business plan to the new realities. The conditions stipulated in the contract, however, remain unchanged.

At the start the private operator receives a new or rehabilitated system with an initial set of connections, from 200 to 1,500 initially depending on funds available during rehabilitation. The number of connections is meant to ensure that the private operator has sufficient customer base to start financially viable operations. This allows the operators to collect revenues from the very beginning. After this initial start, it is up to the operator to define their expansion strategy to make it a profitable business. Expanding services means a larger customer base and greater possibilities for generating revenue and profits. The contractual arrangement stipulates the payment of a fee to AIAS and to CRA to support their services. The private operator is supposed to pay a fee from the revenue they generate from selling water. These fees are currently not transferred to AIAS. This is because AIAS' current organizational status does not allow the organization to collect payments. In the short term, new regulations concerning AIAS will allow the organizations to issue invoices and manage their finances more independently. The first contracts were all granted with the single water tariff[3] applicable to all systems. The reason to implement a singular tariff was that there was insufficient historical data to be able to generate a more accurate tariff that would reflect the actual cost structure of each individual system. The operators to which the contracts

[3] 18 MZN/m^3, volumetric consumption cost or an average 23 MZN/m^3, for 10 m^3, consumption (domestic use) – including a MZN 50 fixed charge per month.

were granted initially had some experience in the water sector but none of them had hitherto operated a water system and lacked knowledge or reference to be able to calculate such a tariff. In early 2015, a first evaluation of prices performed by AIAS with information on operational costs provided by operators revealed that tariffs would have to go up to tariffs 30–32 MZN/m^3, in order to allow operators to recover operational costs[4], fee payment and generate profit. The new tariff required has been negotiated and approved by CRA in 2015. However, until today these revised tariffs have not been implemented. CRA had initially indicated that the tariffs will have to be set according to affordability studies and the purchasing power of the users in these locations. Raising the tariffs remains one of the focus strategies of AIAS as it would allow them to co-fund their operations from the receipt of fees.

AIAS came to realize that in order for their ideal model to work they had to invest in the development of a strong local private sector dedicated to water services. This private sector previously did not exist in Mozambique. Therefore, an important part of the approach of AIAS, which is not formally documented in the contract, concerns capacity development activities and the strengthening of private operators and their staff. Several external organizations implement on-the-job training for operations and support the efforts of reporting on operations. As part of this approach the systems are regularly visited by AIAS staff, or donor staff reporting to AIAS. Each town is recipient of a capacity building activity (i.e. maintenance training, installation of meters, network design) on location. Apart from that, all system operators participate in financial seminars in which information is shared. Other meetings between AIAS and the operator occur whenever there is a need. This approach has grown to include the procurement of materials such as pipes or water meters on behalf of private operators in order to facilitate the expansion of services and increase coverage in small towns.

5. The PPP contract

5.1. What (standards) and where (beneficiaries)

Although there is some variation[5], each contract specifies a standard degree of operational performance in the form of commercial and physical leakages that the operator should achieve, as well as service continuity, which is specified in days of service and hours of service per day. As a general rule the operator should be able to offer service seven days a week for at least 14 hours per day. This, as mentioned, varies per contract and system. The contract does not specify equipment or material standards that are to be used in achieving these service outcomes. The only specifications that exist in the contract make reference to the limits of infrastructure development. As the private operator has an incentive to expand the customer base, the private operator tends to be focused on expanding the network to add new connections. Expansion of services to additional customers, however, may go at the cost of the level of service delivered to existing customers. The expansion of the network may be beyond the

---

[4] AIAS trusts the information generated by the operators, but has not been able to verify whether the reported operational costs are solely operational or if they include as well capital expenditures operators have incurred in order to expand network or other major repairs they have funded themselves.

[5] All contracts have the same structure and sections. However, some contracts are more specific than others in determining which service levels are to be reached. Operators that have been engaged longer with AIAS have more detailed contracts than others.

actual design capacity of the system. Even though this means that active connections would receive less water, having more connections is attractive for the private operator. Operators thus tend to prioritise coverage rates over service levels as more connections potentially enhances revenue generation.

Operators receive a fully rehabilitated system. Any major modification to the system is, similar to other PPP contracts of the same type, the responsibility of the asset holder (AIAS). In our case study towns, no modifications were made to the production systems. However, in expanding the distribution networks instances have been reported[6] in which the operators compromise on the quality of material used for secondary or tertiary network expansion if they see that the costs of extending services cannot be fully covered by the new connections. The contract does not contain specific definitions of customer segments or areas that the operator is to cover. As a result, the operators have reported to selectively expand to those neighborhoods where densities are higher allowing for them to generate additional revenues for relatively little investment as less infrastructure is required to extend services in these more densely populated areas. This is understandable given the contract conditions and incentive structure of the private operator. There are also instances in which the operators have extended services to less attractive areas at the request of the local authorities. This is, however, not the preference of the operator and is directly linked to the intervention of local authorities. Furthermore, there is no clear definition of the service areas that the operators in an 'AIAS town' is to cover. This means it is not entirely clear what areas the operator would need to expand to and which areas remain under the auspices of local authorities or the existing offices for infrastructure planning of the MPOHRH.

Small towns in Mozambique have a very specific population distribution. In these towns it is easy to identify a small 'cement' (planned or urbanized) city center that is surrounded at the fringes by rural and more dispersed development (Raimundo, 2015). The local authorities do not always have the capacity to define or to execute long-term urban plans. This means that development of these towns is largely unplanned. As a result, operators often lack information as to what would be the most reasonable expansion work to provide water, particularly in the context of future development of the small town. As a result, expansions are done as a response to users' demands and on an ad hoc basis rather than in anticipation of such demand. Moreover, the ability to meet user demands is largely dependent on the capacity and ability of the operator to provide for service extensions at the time of the request.

5.2. How many (connections) and when (contract timeframe)

The contract does stipulate a gradual growth in coverage. Since additional revenues are generated mainly through service expansion, this stipulation is not seen as a limitation to the operator. Operators report different break even points for each system. This will depend on the type of treatment system and water source exploited (surface water and contaminated groundwater will generally become more expensive to source because of treatment costs). Initial estimates put the breakeven point at 1,000 connections, which resembles the number of initial active connections AIAS tries to facilitate for the operators[7]. However, due to the 2015 economic crisis in Mozambique and the subsequent inflation

[6] In Caia and Bilene, this was reported to be the case.

[7] There are no studies available that have been able to define a universal break-even point for these systems in this context (Kingdom, 2005).

rates above 25%, the accuracy of this estimate is increasingly questioned. In 2017 one of the operators with AIAS that is currently managing seven systems reported a breakeven point at 2000 connections.

Given that the operator is motivated to expand services to increase revenues and the limited length of the PPP contract of seven years, operators try to accelerate whenever possible by frontloading the expansion of the network with household connections in the first years. In Bilene, coverage has increased by 63% in two years since the contract was signed. This has also happened in Caia (47% total increase), in Moamba (24%) and in Mopeia (33%) This is favorable for the operators because they can increase revenues.

As asset holder, AIAS is responsible, and supports, the rehabilitation and expansion of intake infrastructure, rehabilitation of energy sources and those system expansions falling outside of the right and responsibility of the operator. Operators are responsible for service expansion, which is essential for them to reach breakeven and start generating a profit. By contract, operators are allowed to develop the primary network up to 500 meters, the secondary network up to 1,000 meters and no limits exist for the development of the tertiary network for the expansion of the system. In order to reach areas that are most attractive, however, operators are developing main network infrastructure that surpasses the limit of 500 meters.

6. Adjusting processes based on written contracts

The cases of small town water systems delegated to the private sector in Mozambique suggest that overall PPP contracts can act as a suitable organizing mechanism for small towns, as they have mainly achieved significant progress in extending coverage. In examining the contracts implemented through AIAS, a few important features stand out. The first relates to the proposition of principal-agent theory that posits that information asymmetry and conflicts are intrinsic elements of a principal-agent relationship. In our cases, however, no such conflict was observed. Important in this respect is that instead of having an agent with very different interests as the principal, the cases suggest that the two parties shared a common goal. The private operator sought to expand services as this would allow the operator to expand their customer base, increase revenue and move closer to breaking even and generating a profit. This is perhaps most aptly illustrated by the frontloading of expansion investments in the towns. AIAS shares with the operators the goal of service expansion. Although AIAS's reason for pursuing service expansion is related to improving water access, rather than profit seeking, the motivations of both parties converge in the same outcome. In such a situation contracts can be seen as an organizing mechanism rather than a tool for conflict resolution where goals are shared and the cooperation will remain' as long as all members of the subsystem can be kept reasonably happy' (Waterman & Meier, 1998: 192).

In addition to the shared goals of the operators and AIAS, what also proved important is the frequent consultation between these operators and AIAS to address contingencies that the contract inadequately covered. Illustrative are the investment made by the operator, which should have been made by the asset holder, and the expansion of the distribution network beyond the officially stated limit of 500 meters. Although the contract foresaw that these activities would be carried out by AIAS, in practice in some cases the operator implemented such expansion activities as they were deemed necessary to allow the operator to operate in a sustainable manner. This highlights that formal contracts are, by themselves, not enough as an organizing mechanism for water service delivery. Rather, relational governance appears to complement formal contracts as an organizing mechanism. In relational-governance 'the enforcement of obligations, promises, and expectations occurs through social processes that promote norms of flexibility, solidarity, and information exchange. Flexibility facilitates adaptation to unforeseeable events. Solidarity promotes a bilateral approach to problem solving, creating a commitment to joint action through mutual adjustment. Information sharing facilitates problem solving and adaptation because parties are willing to share private information with one another' (Poppo & Zenger, 2002: 710).

Having said this, we recognize that the idea of goal alignment and the importance of relational governance in the cases we present should be nuanced. Although the goal is shared, the reasons for pursuing these goals are not. Whereas the interest of the operator most likely lies with profit generation, the interests of AIAS and CRA concern the improvement of water access. These different purposes may still lead to potential conflict in the future. In that sense, the formal contracts, incomplete as they are, remain a necessary tool to guide the behavior of the agent should the goals diverge.

Moreover, it is important to recognize that the existing contracts do show room for improvement. Although operators have the possibility to expand the network and increase the number of connections to increase their revenue, the limited contract time means that these expansions may not materialize in time for the operator to break even. Furthermore, there is no clear specification in the contract defining the service area to be covered by the operator in an AIAS town. One approach would be for AIAS to limit its mandate to expanding services to the most urban areas of these towns. These are arguably the most profitable areas for operators and where infrastructure expansion is more viable. The infrastructure located at the fringes of the town such as (communal) standpipes would then continue to be supported by other organizations.

Apart from these gaps, there are also future concerns, of which the development of the tariff is perhaps the most important as different governmental officials differ in opinion on this topic. The regulator claims that the ability to pay in small towns is limited and that the price of water should reflect that. However, the operators and AIAS both have an interest in a tariff structure that warrants financial sustainability in each and every system. This is necessary in the eyes of the operator as it allows them to effectively run the system, and eventually generate profit. This is also necessary in the eyes of AIAS as it allows them to include a service fee they can use in the common pool fund to further develop their work with other systems supplementing government budgets for infrastructure development. The difficulty to raise tariffs paired with the limited ability to pay, as argued by local government representatives and CRA, may pose issues in the development of business opportunities.

6.1. Enabling conditions

The cases also highlight a number of conditions that have greatly aided the success of the contracts. These enabling conditions are not necessarily a formal element of the contract, but did have an important impact on how the private operator was able to perform showing once again the importance of relational governance over contract restrictions.

6.1.1. Support for the small-scale private sector. The approach to PPP contracts as they are currently being implemented around the globe usually adhere to standards based on, and perhaps achievable in, large urban areas. But the realities in which private operators have to provide services in small towns is very different from the large urban centers. Small towns have a much smaller market from which to generate revenues, lower population densities, and a highly diverse population in terms of household income. Such areas require a different set of 'efficiency' standards and as such as different approach to PPP (Hirn, 2013; Ameyaw Effa *et al.*, 2015). For the private operators in the AIAS towns, much more support is required as compared to traditional approaches to PPP in the water sector. Without such additional support, ranging from capacity strengthening to assisting in procurement of materials and equipment, PPPs in these settlements may not maximize the value of the arrangement for all parties involved and may not incentivize the participation of private companies. As explained by one of the operators: 'Only because we had the initial support of AIAS we initially entered into this contract. We were technically guaranteed that we were not running considerable risks, so we could experiment' (operator, personal communication).

6.1.2. Creating a running start. Important in the PPP contracts in small towns in Mozambique is that the system was fully rehabilitated prior to transfer to the private operator. Moreover, the systems already had a sufficient number of connections[8] that would allow the private operator to break even and start generating a profit within a reasonable time. This was of particular importance given the limited duration of the contracts. The five year contract with a possible extension of two years means that the operator has limited time to invest in service expansion with the aim of recovering costs and achieving a profit within the contract time. By ensuring that a transferred system has between 200 to 1,500 connections AIAS not only ensures that private operators will be able to break even, but also that the private sector will be interested in providing water services (Kleemeier & Lockwood, 2012). Particularly the combination of the government support highlighted above and providing a rehabilitated network with a sufficient number of connections was important in creating private sector interest.

6.1.3. 'Pragmatic' contracts. The contracts implemented in Mozambique by AIAS resemble very much standard contracts implemented elsewhere. Five years down the road, however, AIAS has had sufficient experience and reason to approach these contracts in a 'pragmatic' manner rather than in a dogmatic manner. In other words, the implementation of the contracts has been relatively flexible, allowing for points of conflict to be avoided and even mitigated. Allowing a certain degree of flexibility to adjust to the specific economic, political and social context in which such a practice is implemented has been recommended before, elsewhere (Aoki, 2000: 17). Local conditions determine to what extent and how a model such as PPP contracts for small-scale private operators can be implemented. These conditions concern aspects like the local capacity to plan and support the operator, or the capacity of the market to grow and pay the price of water that operators would require to sustain the system. In our cases, local conditions forced both operators and governmental agencies to adjust the processes initially foreseen in the conditions as the PPP contracts in order to ensure that the interests of both the principal and the agent have been addressed.

7. Conclusion: reconsidering principal-agent relationships

The model of PPP involving small-scale private operators has increasingly been promoted in recent years. Underlying this model is the use of formal contracts as an organizing mechanism for the delivery of water services in small towns. The success of models for water services is often attributed to a willing provider, a supportive government and a comprehensive set of regulations (Ndaw, 2016). As we have

[8] Except in Moamba where the system was transferred with zero connections.

seen in Mozambique, the existence of these elements is important, but in no way sufficient to lead to a successful implementation of the model. The implementation of the model in towns that we studied suggests that it can work. At the same time, the explanations of why it has worked seem to relate less to the actual model and more the adjustments that occurred during the implementation.

Key in understanding the implementation of the model is the delicate balance between and complementarity of the role of formal contracts as a mechanism to shape behavior and relational governance as a means to address the gaps and obstacles that come forth from formal contracts. In a way they represent two sides of the same coin and become important at different stages of the implementation of contracts. Crucial in this respect is also the alignment of the objectives of the agent and the principal. Given that both parties understand the shared benefit from service expansion, we argue that the relevance of relational governance has become more important in understanding the relationship between the parties, rather than a literal study and interpretation of the formal contract. It remains to be seen if relational governance would have the same prominence if the interests of the agent and the principal would have diverted. However, based on the current development we would strongly propose a revision of the current understanding of contracts as means of control mechanisms of the principal over the agent.

Moreover, the actual implementation of the PPP contracts in Mozambique had a few very important adjustments that allowed it to function in the specific context of small towns in Mozambique. The support to private operators, the pragmatic interpretation of the contract, the emphasis of creating a running start for the private operator were all crucial in ensuring a successful first stage for the implementation of this model. The main lesson learnt is that the question is not so much about the model and if it works. The question is more about how to change, adapt and adjust the model to let it work in the institutional, demographic, socio-economic and bio-physical conditions of small towns in Mozambique. In order to allow for these adjustments, parties need to be willing to reinterpret the interests, incentives, possibilities and limitations of all parties involved.

Having said this, questions do remain about the future of the PPP model in the small towns of Mozambique. Questions remain surrounding the development of suitable tariff levels that serve all parties, operator, asset holder and regulator (and households), which service areas and locations of small towns that should be served by small-scale private operators. Moreover, not all operators holding a contract with AIAS have the same capacities (technical, financial and organizational). The end of the contractual period with those operators who signed their contracts in 2014 is closely approaching. This offers an opportunity to revise the relationships of the actors involved and the conditions that govern this relationship.

References

Adank, M. (2013). Small town water services: Trends, challenges and models. (Thematic Overview Paper 27). IRC International Water and Sanitation Centre, The Hague.

Adinyira, E., Sohail, M., Oteng-Seifah, S. & Adjei-Kum, T. (2010). Private sector driven supply chains in Ghana's rural water sector. *Water Policy. 12*(3), 416–424.

Ameyaw Effa, E., Chan, A. P. C. & Owusu-Manu, D. G. (2015). Domestic private sector participation in small-town water supply services in Ghana: reflections on experience and policy implications. *Public Organization Review 15*, 175–192.

Annez, P. C. (2006). *Urban Infrastructure Finance from Private Operators: What have we learned from recent experience?* (Vol. 4045). World Bank Publications, Washington, DC.

Aoki, M. (2000). Institutional evolution as punctuated equilibria. In: *Institutions, Contracts and Organizations: Perspectives From New Institutional Economics*. Menard, C. (ed.). Edward Elgar, Northampton, MA, pp. 11–35.

Cohen, B. (2006). Urbanization in developing countries: Current trends, future projections, and key challenges for sustainability. *Technology in Society 28*, 63–80.

Cooper, P. (2003). *Governing by Contract: Challenges and Opportunities for Public Managers*. CQ Press, Washington.

Delmon, V. (2012). *Public Private Partnerships: Contracts*. Training Session World Bank, Dakar, Senegal. June.

Ferre, C., Ferreira, F. & Lanjouw, P. (2012). Is There a metropolitan bias? The relationship between poverty and city size in a selection of developing countries. *The World Bank Economic Review 26*(3), 351–382.

Gleick, P. H. (1998). The human right to water. *Water Policy 1*(5), 487–503.

Harvey, P. & Reed, R. (2006). Sustainable supply chains for rural water supplies in Africa. *Proceedings of the ICE-Engineering Sustainability 159*(1), 31–39.

Harvey, P. A. & Reed, R. A. (2007). Community-managed water supplies in Africa: sustainable or dispensable? *Community Development Journal 42*(3), 365–378.

Hirn, M. (2013). Private Sector Participation in the Ugandan Water Sector: A review of 10 Years of Private Management of Small Town Water Systems. In: *Sustainable Services Through Domestic Private Sector Participation*. Water and Sanitation Program: Working Paper, Washington, DC.

Hopkins, R. (2003). Town Water Supply and Sanitation Services Volume 3: An Alternative Perspective on WSS Services (including the 'Grey Area'). Companion Paper A1. The World Bank, Washington, DC.

Kariuki, M., Schwartz, J. & Schur, M. (2006). Reaching Unserved Communities in Africa with Basic Services: Can Small-Scale Private Service Providers Save the Day?

Kingdom, W. (2005). Models of aggregation for water and sanitation provision. Water Supply and Sanitation Working Notes. Note No. 1, January 2005. World Bank, Washington, DC.

Kleemeier, E. L. (2010). Private operators and rural water supplies: A desk review of experience.

Kleemeier, E. & Lockwood, H. (2012). *Public-private Partnerships for Rural Water Services*. IRC International Water and Sanitation Centre, The Netherlands.

Komives, K. (2001). Designing pro-poor water and sewer concessions: early lessons from Bolivia. *Water Policy 3*, 61–79.

Lane, J. (2000). *New Public Management*. Routledge, London.

Lane, J. (2001). From long-term contracting to short-term contracting. *Public Administration 79*(1), 29–47.

Larbi, E. (2005). *Public-private Partnerships and the Poor in Water Supply Projects: The Ghanaian Experience*. Well Factsheet, Ghana.

Lockwood, H. & Le Gouais, A. (2011). *Professionalising Community-Based Management for Rural Water Services*. IRC International Water and Sanitation Centre, The Hague.

Mara, D. & Alabaster, G. (2008). A new paradigm for low-cost urban water supplies and sanitation in developing countries. *Water Policy 10*, 119–129.

Mehta, M., Virjee, K. & Njoroge, S. (2007). Helping a New Breed of Private Water Operators Access Infrastructure Finance: Microfinance for Community Water Schemes in Kenya.

Moriarty, P., Patricot, G., Bastemeijer, T., Smet, J. & Van der Voorden, C. (2002). *Between Rural and Urban: Towards Sustainable Management of Water Supply Systems in Small Towns in Africa*. IRC International Water and Sanitation Centre. Delft, The Netherlands.

Moriarty, P., Smits, S., Butterworth, J. & Franceys, R. (2013). Trends in rural water supply: towards a service delivery approach. *Water Alternatives 6*(3), 329–349.

Mugabi, J. & Njiru, C. (2006). Managing water services in small towns: challenges and reform issues for low-income countries. *Journal of Urban Planning and Development 132*(4), 187–192.

Ndaw, M. F. (2016). *Private Sector Provision of Water Supply and Sanitation Services in Rural Areas and Small Towns: The Role of the Public Sector*. Water and Sanitation Program: Guidance Note.

Pilgrim, N., Roche, B., Kalbermatten, J., Revels, C. & Kariuki, M. (2007). Principles of Town Water Supply and Sanitation. Part 1: Water Supply. World Bank Water Working Notes, Washington, DC.

Poppo, L. & Zenger, T. (2002). Do formal contracts and relational governance function as substitutes or complements? *Strategic Management Journal 23*(8), 707–725.

Prasad, N. (2006). Privatisation results: private sector participation in water services after 15 years. *Development Policy Review 24*(6), 669–692.

Raimundo, I. (2015). Urbanisation of African cities: urban patterns in small- and medium-sized cities of Mozambique (unpublished).

Rondinelli, D. (1983). Towns and small cities in developing countries. *Geographical Review 73*(4), 379–395.

Schaub-Jones, D. (2008). Harnessing entrepreneurship in the water sector: Expanding water services through independent network operators. *Waterlines 27*(4), 270–288.

Schwartz, K., Tutusaus, M. & Savelli, E. (2017). Water for the urban poor: balancing financial and social objectives through service differentiation in the Kenyan water sector. *Utilities Policy 48*, 21–33.

Vincent-Jones, P. (2000). Contractual governance: institutional and organizational analysis. *Oxford Journal of Legal Studies 20*(3), 317–351.

Water and Sanitation Program (2010). *Sustainable Management of Small Water Supply Systems in Africa*. Practitioners' Workshop Report, October 6-8 2010. Field Note. World Bank.

Waterman, R. & Meier, K. (1998). Principal-Agent models: an expansion? *Journal of Public Administration Research and Theory 2*, 173–202.

WHO (2015). 25 Years: Progress on Sanitation and Drinking Water – 2015 Update and MDG Assessment. UNICEF and World Health Organization, New York.

World Bank (1994). World Development Report 1994: Infrastructure for Development. Oxford University Press, New York.

WSP-AIAS (2015). Documento de Apresentacao da AIAS: 2009–2014. 5 Anos Elevando a Qualidade de Vida das Populacoes. Maio 2015. Maputo (unpublished).

World Bank (2014). *Water PPPs in Africa*. World Bank Group.

Small town water services sustainability checks: development and application in Ethiopia

Marieke Adank[a], Sam Godfrey[b], John Butterworth[c] and Eyob Defere[c]

[a]*Corresponding author. IRC, P.O. Box 82327, The Hague 2508 EH, The Netherlands. E-mail: adank@ircwash.org*
[b]*UNICEF, UNECA Compound, NOF-Building #020, Addis Ababa 1169, Ethiopia*
[c]*IRC Ethiopia, P.O. Box 2, Addis Ababa 1251, Ethiopia*

Abstract

With rising coverage figures and the advent of the Sustainable Development Goals, there is increasing attention given to assessing and monitoring the sustainability of water services. Previous efforts in the rural water supply sector have included the development of sustainability checks, while in the urban water supply sector, benchmarking of water services and the performance of utilities has become common practice. This paper argues that neither rural sustainability checks, nor urban benchmarking frameworks, are entirely suitable for monitoring small town water services. It presents a framework specifically developed and applied for assessing and monitoring small town water services. Application of the framework in seven small towns in Ethiopia shows significant discrepancies between the ideal and actual situations. It reveals specific challenges related to sustainable small town water service provision, including capacity at service provider (utility) level, asset management and regulation. The costs of sustainability checks and prospects for uptake as project and wider sector tools are discussed.

Keywords: Ethiopia; Monitoring; Small town; Sustainability check; Water services

1. Introduction

The Millennium Development Goal (MDG) 7 target C on halving the proportion of people without access to safe drinking water was met in 2010 (WHO/UNICEF, 2014). However, doubts have been expressed about the quality of the 'access to safe water' and whether this achievement is telling the full story. Taking into account water quality, for example, significantly reduces coverage figures, as shown by various authors (Godfrey & Labhestwar, 2010; Bain *et al.*, 2012; Onda *et al.*, 2012). These concerns have been driving the development of indicators for measuring water services, going beyond simply counting facilities and tracking 'the haves and the have-nots' (Bartram, 2008; Lockwood & Smits, 2011) and combining indicators to assess the level of services provided and received (Kayser

doi: 10.2166/wp.2018.004

et al., 2013). The recent advent of the Sustainable Development Goals (SDGs) has emphasised the need for comprehensive monitoring frameworks that integrate quality, affordability, continuity and equitability of water, sanitation and hygiene (WASH) services provision.

At the same time, there is increasing interest in finding ways to assess whether or not the conditions for sustainable water service provision are in place. Development partners such as DGIS (Netherlands Directorate-General for International Cooperation), USAID, and more recently, DfID (ICAI, 2016) have been stimulating the development and application of so-called 'sustainability checks'. These aim to assess whether their investments have led to continuously functioning systems, and the degree to which conditions for future sustainable WASH service provision are in place. Sustainability checks were first developed and implemented in UNICEF Mozambique's 'One Million Initiative' programme (2007–2013) with financial support from DGIS (Godfrey *et al.*, 2015). Subsequently, sustainability checks have been systematically applied in UNICEF programmes funded by DGIS (Boulenouar, 2016). Other frameworks for assessing sustainability which have been developed over the last decade include USAID's Sustainability Index Tool or SIT (USAID, 2013), and the WASH Alliance International's Sustainability Monitoring Framework (Dutch WASH Alliance, undated). The majority of these project-based sustainability check frameworks have focused on rural water (and sanitation) service provision, focussed on point sources (Boulenouar *et al.*, 2013).

In the urban water sector, performance benchmarking has been gaining popularity since the 1990s and is now being applied worldwide (Murungi & Blokland, 2016). This kind of benchmarking also looks beyond coverage and commonly includes sustainability indicators, especially related to operational efficiency, financial efficiency, production and consumption and poverty and affordability. Since 1996 IBNET (International Benchmarking Network for Water and Sanitation Utilities) has been providing standardised measurements of utilities' operational and financial performance. It provides access to information on the performance of more than 2,500 water and sanitation utilities from 110 countries, including Ethiopia (van den Berg & Danilenko, 2010).

1.1. Small town sustainability checks

In line with the topic of this Special Issue, this paper focuses on assessing and monitoring water services in small towns. Small town water services provision spans the middle ground of the rural-urban continuum: not fully urban, as it usually does not concern big public utilities managing complex piped schemes, and not fully rural, as the common rural model of water service provision through community-managed handpumps is not a realistic model in these settings either (Adank, 2013; Moriarty *et al.*, 2002). Various service delivery models can be found in small town settings, including municipal-management, community-management, public-owner operators, and public-private partnership arrangements. The type of management model commonly depends on the local context and the complexity of the technical scheme (Adank, 2013; Pilgram *et al.*, 2007). The diversity of service delivery models and the mix of urban and rural characteristics limits the applicability of strictly urban or rural monitoring frameworks.

As mentioned above, rural water sustainability checks tend to focus on one specific service delivery model, community-managed handpumps, which is less relevant in small town settings. Urban benchmarking frameworks are too complex, too comprehensive, or not fully relevant for application in small town contexts. The data needed in order to do this kind of benchmarking is often not available in small towns. Since aggregation of data is an important requirement for the authorities overseeing small town water services provision, monitoring tools need to be designed with this in mind.

Governance and legitimacy issues and issues related to the enabling environment are also often not included into utility performance benchmarking. Where urban utilities generally do not require technical support from the public sector, as they either have this in-house or access the private sector for such support, small town water service providers generally do require direct support. The monitoring of such support functions is critical for small town water service provision.

Small towns have a unique institutional arrangement. The nature of the small town population precludes the use of 'community management' of the piped water service but often does not reach the level of 'professional water service delivery' that is achieved in the capital or large cities. The lack of professionalised water utilities in many small towns in Africa, precludes the use of utility benchmarking as a means of measuring water service sustainability. In this context, how can small town water service provision be checked and monitored? What indicators are most useful for monitoring small town water service provision and what conditions need to be in place? What do results of monitoring small town water services tell us and how can these results be used? This paper seeks to provide answers to these questions by presenting and discussing the case of the development of a sustainability check framework for small and medium town WASH services provision and its application in seven towns in Ethiopia.

2. Setting the scene: small town water services provision in Ethiopia

Ethiopia is a mostly rural country undergoing rapid urbanisation. While around 80% of the population live in settlements with fewer than 2000 people (CSA, 2013) there is considerable urban growth. The number of people living in towns and cities is expected to increase from 15.2 million in 2012 to 42.3 million by 2037, increasing at 3.8% per year (CSA cited by World Bank, 2015), or may even triple by 2034 based on the World Bank's growth rate projection (5.4% a year).

With the exception of two chartered cities (Addis Ababa and Dire Dawa), which form their own administrative states, every town or city is part of a regional state. As a federal republic, Ethiopia is divided into nine national regional states, mainly based on ethnicity and language. Regions are subdivided into '*woredas*' (districts) which are further subdivided into '*kebeles*' (wards or neighbourhood associations). Woredas are grouped into zones, which either have a role in supervising the woredas (like in in Oromia and Tigray Regions) or play a less active role (like in Amhara and Somali Regions). Towns with a population of at least 17,500 have a mayor and a town council (Ministry of Urban Development & Housing, 2013) and tend to form an '*urban woreda*' or municipality, while smaller towns tend to be part of a '*rural woreda*' and fall administratively under a woreda council.

Every settlement with a population above 2,000 and any settlement which serves as a woreda capital is considered a 'town' in Ethiopia. Out of the 882 urban settlements counted by the Ministry of Water, Irrigation and Energy (MoWIE) in 2014, the vast majority (847, which amounts to 96%) were small (2000–20,000 population) and medium (20,000–50,000) size towns (MoWIE, 2014). These small and medium towns are considered strategic for development due to their rapid population growth and their importance as centres of local business and growth within their rural hinterlands. The current lack of infrastructure and services in such towns is identified as a critical gap and a risk to the success of urbanisation policies (World Bank, 2015).

According to WHO/UNICEF's Joint Monitoring Programme (JMP), access to improved water services in urban areas in Ethiopia has improved from 84% in 1990 to 93% in 2015 and access to piped water on premises also increased considerably, from 10% to 56% during the same time period (WHO/UNICEF, 2017).

However, these largely positive coverage figures hide the reality of low levels of service and challenges related to the sustainable provision of small town water services (Adank *et al.*, 2016).

Small town water services provision is, in Ethiopia, a part of urban water service provision. Service delivery models applied in small towns do not differ considerably from service delivery models applied in cities. Small town piped schemes are commonly managed by Town Water Utilities with a Water Board. These schemes provide services through household connections and public standpipes. Regional Water Bureaux are responsible for the development of such schemes and they tend to contract out activities like drilling, construction, capacity building and supervision to private sector enterprises. As soon as the infrastructure is put in place, it is handed over to the town administration, which is then responsible for establishing a Town Water Supply and Sewerage Service, also referred to as a Town Water Utility. These autonomous public enterprises are responsible for the management of such piped schemes. The general manager of a Town Water Utility is responsible for organising, directing and administering the activities of the utility and its staff within the different sections (human resources development, finance and property administration, operation and maintenance, etc.). The Town Water Utility is overseen by a Water Board, which is supposed to lead and regulate the service as a supervisory body on behalf of the town council and other represented stakeholders. It also appoints the general manager of the Town Water Utility, who is a non-voting member of the Water Board. In addition, Town Water Boards generally consist of representatives from the town administration, from the relevant local government offices (e.g. water office, health office, finance and economic development office, education office), and from the customers (a male and a female representative from the domestic customers and possibly a representative from the business community) (MoWIE, 2013). Water Board members are to meet on a regular basis. Sitting allowances related to such meetings are generally covered by the Town Water Utility through revenues from tariffs.

Town Water Boards are established based on regional proclamations. Water Boards of small towns (category 3–5, with a population of less than 100,000) are established by the town administration and are hence officially accountable to the town administration[1] (MoWIE, 2013). However, in reality there are no performance agreements between a Water Board and a town administration and therefore limited formal accountability between the two. The main role the town administration plays is approving tariffs and in providing match-funding in case of rehabilitation or expansion of the scheme.

The Regional Water Bureau is to continuously follow up on the performance of both Town Water Boards and Town Water Utilities, and to give training and technical support when needed (MoWIE, 2013).

3. A sustainability check for small town water service provision: framework and methodology

A sustainability check framework and methodology was developed as part of the ONEWASH Plus Programme[2]. This programme, implemented by UNICEF in collaboration with the Government of Ethiopia and with financial support from DfID, focuses on improving sustainable WASH services in

---

[1] In Oromia Region, Water Boards can also be established by the Zonal Water Bureau. Water Boards for towns with a population of 100,000 or more (grade 1 and 2) are established by the Regional Water Bureau.

[2] The three authors of this paper from IRC have been actively involved in the monitoring and evaluation component of this programme, including the development of the sustainability check framework.

eight small towns and their surrounding rural areas. The development of the framework drew on existing frameworks and experiences with urban benchmarking and mostly rural-focused sustainability checks presented in the introduction. Based on these frameworks, and taking into account suggested indicators, norms and standards as set out in Ethiopia's One WASH National Programme (OWNP) (Federal Democratic Republic of Ethiopia, 2013) and the second Growth and Transformation Plan (Federal Democratic Republic of Ethiopia, 2015), a draft sustainability framework was developed to suit the Ethiopian small town context. The draft framework was discussed with stakeholders during a series of consultation workshops in 2015 at town, regional and national level, including representatives from government, NGOs and development partners. Feedback from the consultations at these levels was used to further refine the framework.

The check was first applied in 2015 in seven small towns[3] and is subsequently being applied on an annual basis over the duration of the project (2014–2018). They continue to provide the project with insights into what is going well and what needs additional attention, including whether or not the necessary conditions for sustainable WASH service provision are in place. In addition, results are used as an input for the development of town-level sustainability plans. In this way, the results are intended to support decision making at town (utility) level related to planning and corrective actions.

The ambition of the development and application of sustainability checks goes beyond the project. The initiative intends to pilot useful indicators and mechanisms for monitoring and regulating small town WASH which could be applied at scale within Ethiopia's sector-wide One WASH National Programme.

3.1. Service levels and sustainability indicators

The sustainability check framework developed under the ONEWASH Plus Programme consists of MS Excel sheets with indicators and scoring tables. This includes (1) indicators related to the level of services provided and (2) indicators for assessing the degree to which the conditions for sustainable WASH service provision are in place. The framework has separate indicator sets for small town water, small town sanitation, rural water, rural sanitation, and institutional WASH. The rural component was included as the programme also targets satellite villages around the towns. This paper focuses on the indicators related to small town water services provision.

The indicators related to level of service were set in line with the minimum service levels as defined in Ethiopia's second Growth and Transformation Plan (GTP-2) and summarised in Table 1. These are a step up from the minimum urban water supply access standard level under GTP-1 which was set at 20 litres per capita per day within a distance of 500 m for urban people. The goal of GTP-2 (2016–2020) is to ensure access for everyone at the GTP-1 level, while 75% of the urban population are expected to receive services at the improved GTP-2 levels (FDRE, 2015).

The sustainability indicator framework follows a structure which is commonly applied in rural sustainability check methodologies, where indicators are defined to assess the presence of conditions for sustainable WASH service provision related to institutional/managerial, technical, financial, environmental and social sustainability (also referred to as FIETS: Financial, Institutional, Environmental,

[3] As implementation in the project town Jijiga focuses on solid waste management, the water component of the sustainability check was not applied in this town.

Table 1. GTP-2 normative service levels.

Town category	Population	Water quantity	Water quality	Accessibility	Reliability
Category 5 (small)	<20,000	40 lpcd	In line with water quality standards of WHO, supplied by schemes labelled as 'improved' by Joint Monitoring Program (JMP) of UNICEF and WHO	Within 250 m	Uninterrupted for at least 16 hours per day
Category 4 (medium)	20,000–50,000	50 lpcd			
Category 3	50,001–100,000	60 lpcd			
Category 2	100,001–1 million	80 lpcd			
Category 1	>1 million	100 lpcd			

Source: Compiled from FDRE (2015) by authors.

Technical, Social (Rijksdienst voor Ondernemend Nederland RVO.nl, 2016)) at different institutional levels. The sustainability check framework developed under the ONEWASH Plus Programme considers three institutional levels: the service provider level, the service authority level and the national level. Functions at the service provider level are related to the day-to-day operation and maintenance of water schemes. Service authority functions include strategic planning, oversight of the service providers, provision of technical support to service providers and monitoring (Lockwood & Smits, 2011). In small towns in Ethiopia, service provision functions clearly lie with the Town Water Utility as, while service authority functions are divided over three entities at different levels: the Water Board, which oversees the service provider functions of the Town Water Utility, the town administration, to whom the Water Board is accountable, and the Regional Water Bureau, who oversee and provide technical support to the Water Boards and Town Water Utilities.

As such, the service authority level determines the local enabling environment in which service providers operate. The enabling environment at national level sets the policy and normative frameworks, as well as the systems and mechanisms for sector monitoring, regulation and capacity development.

Table 2 provides an overview of the indicators at the three levels. Several of these indicators build on urban water service benchmarking indicators (e.g. staff productivity and billing and collection).

3.2. Scoring tables

A scoring system was developed with micro-scenarios to which ordinal scores were attached, going from 0 to 100, with a 50 score indicating the minimal acceptable score (the benchmark). For the service level indicators, the 50 score were set to equal the GTP-2 norm. For the sustainability indicators scores have been set as much as possible in line with national guidelines and in consultation with town, regional and national stakeholders. The draft framework and scoring system were discussed with a wide variety of stakeholders, including representatives from government, NGOs and development partners, through consultation workshops at different institutional levels: national, regional and town level. Feedback from the consultations at the three levels was used to further refine the framework and resulted in the modification, deletion and addition of some indicators and scoring elements. An overview of the indicators, micro-scenarios and the associated scores can be found in Annex 1 (available with the online version of this paper).

It is sometimes argued that ordinal variables cannot be treated as numeric, which means they cannot be added, subtracted, divided and multiplied, as the distance between the different steps between ordinal

Table 2. Summary of indicators included in the small town water services provision sustainability check.

Service provider	Service authority	National enabling environment
Institutional sustainability		
SP-I-1: Effective utility management SP-I-2: Staff productivity SP-I-3: TWU staffing	SA-I-1: Sufficient capacity at regional and zonal level to provide support to TWUs SA-I-2: Presence of Water Board (WB)	N-I-1: National monitoring system or database N-I-2: National support to Regional, Zonal and Woreda/service authority, including capacity building training and technical backup N-I-3: Urban regulatory agency
Technical sustainability		
SP-T-1: Non-revenue water SP-T-2: Adequate supply of spare parts for minor maintenance (pipes, fittings etc.) SP-T-3: Effective maintenance system in place SP-T-4: Water quality management and disinfestation	SA-T-1: Effective provision of technical support to the TWU SA-T-2: Checks on construction quality	N-T-1: National/regional standard/ guideline/ norms for urban water service provision
Financial sustainability		
SP-F-1: Cost recovery SP-F-2: Effective financial management SP-F-3: Effective asset management SP-F-4: Effective billing and collection		N-F-1: Availability of national budget related to urban water supply
Environmental sustainability		
	SA-E-1: Catchment management system in place	N-E-1: National environmental protection standards are established and applied to town water services
Social sustainability		
SP-S-1: Urban poor get affordable water		

Source: Authors.

scales are not exactly the same (Pelto & Pelto, 1978). However, according to Wijk-Sijbesma (2001), if properly constructed these kinds of scales can be treated as interval scales with an assumed equal distance between subsequent steps. Care was taken to construct the scoring tables related to each indicator ensuring more or less equal distance between the scores. This allows aggregation of scores through summation or averaging of scores over multiple indicators or multiple units of observation, which will be especially relevant if the framework is taken to scale.

3.3. Data collection and analysis

Different data collection methods were applied to collect the data needed for the scoring of the service level and sustainability indicators. Some of the service level indicators, like the proportion of households within 250 metres, can only be sensibly scored based on household data, while others can be scored based on observations, available secondary data at town level (e.g. financial records), or on information from key informants.

In the case presented here, data on service levels were collected as part of the ONEWASH Plus baseline data collection exercise. This included a household survey administered to 100 randomly selected households in each town between 22 September and 12 December 2014 by local enumerators supervised by IRC. The survey included questions about the accessibility of the water scheme and the payment modalities for water services. In addition, a water point survey was conducted at all public standpipes in the seven towns. Data were collected by enumerators using surveys loaded on phones using Akvo FLOW software. Further methodological details on collection of service level data are included in Adank *et al.* (2016).

The data required for scoring of the sustainability indicators at service providers, service authorities and the national enabling environment level were mostly collected through key informant interviews with staff from the Regional Water Bureaux and the Town Water Utility, and undertaken by experienced Ethiopian WASH experts in June 2015.

Data were used by project staff to determine scores associated with the service level, service provider, service authority and national enabling environment indicators. The findings were presented to town and regional level stakeholders for validation and, where needed, correction in December 2015.

3.4. Introduction to the project towns

The seven ONEWASH Plus project towns considered in this paper are served by piped schemes providing services through private connections and public standpipes. The towns were selected by the Government of Ethiopia and UNICEF in 2013 based on access criteria. Six of the seven ONEWASH Plus project towns are served by a piped scheme supplied by one or multiple boreholes, with water treatment limited to disinfection. The piped scheme of Adishihu is supplied by a protected spring. Sheno and Wukro have a relatively high number of household connections and Welenchiti the highest number of public standpipes. The piped schemes are managed by Town Water Utilities, each with a Water Board, set up in line with Regional proclamations (Table 3).

In five of the seven towns, more than 90% of people rely on water from the piped scheme through either household connections ('piped on premise') or public taps or standpipes (Figure 1 and Table 4). In Kebridehar, the majority of people (66%) depend on birkas or other household level storage facilities, commonly filled by tanker trucks or cart vendors. In Maksegnit, about a third of households depend on alternative improved sources such as communal handpumps in addition to water supply from the piped scheme.

4. Applying the sustainability check framework: results

4.1. Small town water service levels

The sustainability check results showed that the seven Town Water Utilities struggle to provide water services as per national norms and standards (Table 1). Achieving reliability and water supply quantity targets were particular challenges, as shown in Table 4. Only the amount of water sold in Welenchiti reached at least 75% of the GTP-2 norm. In all towns water supply is intermittent, with rotation of

Table 3. Overview of pilot small towns, ordered from smallest to largest population size.

Town	Adishihu	Sheno	Maksegnit	Abomsa	Welenchiti	Kebridehar	Wukro
Region	Tigray	Oromia	Amhara	Oromia	Oromia	Somali	Tigray
Town population, 2014 (CSA, 2013)	10,771	15,459	16,930	20,517	21,282	35,807	49,925
Year of construction of piped scheme	1996	1985	1996	1994	1970	1978	1986
Year of latest rehabilitation of piped scheme	NA	2011	2013	2014	NA	2013	2004
Number of functional sources	1spring	3BHs	1 BH	2BHs	5BHs	2BHs	6BHs
Electricity supply	National grid + generator	National grid ± generator	National grid	National grid	National grid	Generator	National grid + generator
Number of household connections, 2014	687	2078	823	1928	1673	1000	5147
Number of public water points, 2014	11	11	10	29	38	22	4

BH = borehole.
Source: Authors.

supply to service areas in different parts of the town at least for part of the year. None of the towns meet the GTP-2 norm of having at least 16 hours of uninterrupted water supply per day. Water quality is also problematic in many towns. Only in Sheno and Wukro were all water quality samples from the piped scheme found to be safe (with an *Escherichia coli* count of less than 10 MPN per 100 ml). The accessibility score is closely related to the number of household connections: the two towns with the most

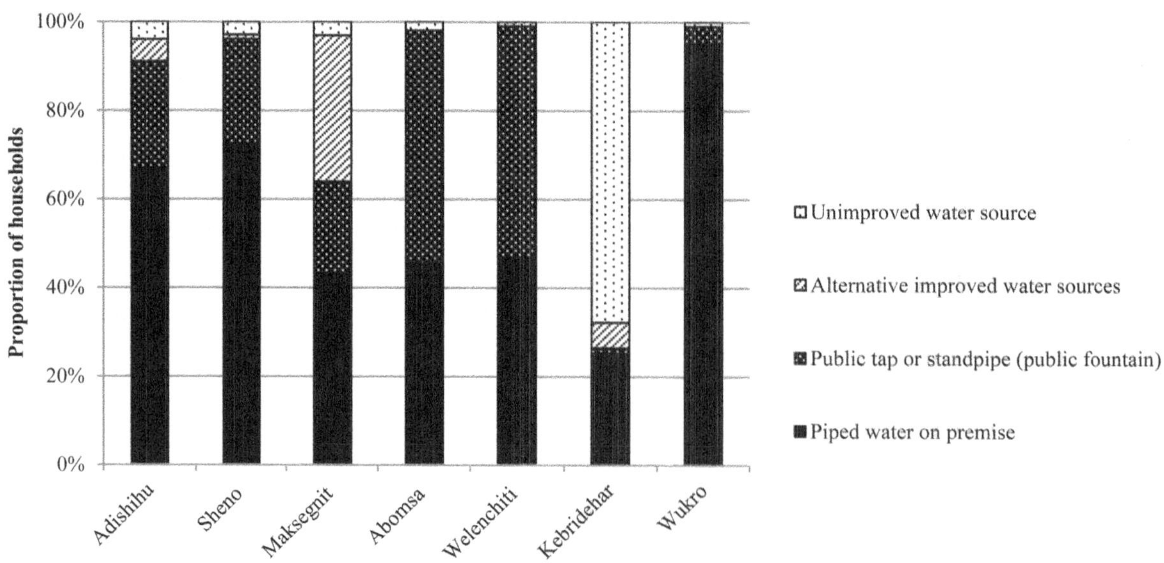

Fig. 1. Sources of main dry season water supply in ONEWASH Plus towns. *Source:* Authors.

Table 4. Service level scores for ONEWASH Plus Programme towns.

	Adishihu	Sheno	Maksegnit	Abomsa	Welenchiti	Kebridehar	Wukro
Piped scheme coverage: % of households which use piped scheme as their main source of dry season water supply, based on household survey	91%	96%	64%	98%	99%	27%	99%
Reliability score	0	0	0	0	0	0	0
Accessibility score	50	50	25	25	25	0	75
Quality score	25	50	0	25	0	25	50
Quantity score	25	25	0	0	25	0	0
Average score	25	31	0	12.5	12.5	6	31
Number of service level benchmarks met	1	2	0	0	1	0	2

Source: Authors.

household connections (Wukro and Sheno) scored highest and the towns with the lowest numbers of household connections (Maksegnit and Kebridehar) scored lowest.

4.2. Small town sustainability indicator scores

Table 5 presents the scores on the different sustainability indicators at service provider level. It shows widely varying scores over the different indicators and the different towns. Overall Wukro scores best and Kebridehar worst on these indicators.

Table 6 presents the indicator scores at service authority level. With the exception of Kebridehar, service authorities met the benchmark on three out of four indicators. The towns in Tigray (Adishihu and Wukro) scored best.

From Table 7 it is clear that there are also considerable challenges in Ethiopia with the enabling environment for small town water service provision at national level. The relatively low scores related to the enabling environment at national level, are not unique to small town (or urban) water supply with rural water supply facing similar challenges.

5. Discussion

5.1. Indicators for monitoring small town service levels

The framework presented in this paper allows for undertaking a rapid assessment of water service levels based on national (GTP-2) norms and standards. In Figure 2 results from the seven towns are compared with the level of service resulting from applying the methodology developed recently by the WHO/UNICEF Joint Monitoring Programme (JMP) for monitoring the Sustainable Development Goals. The JMP analyses nationally representative data, while the sustainability check presented here is mainly intended for assessing water service levels of individual towns, with the possibility of aggregation of scores at higher levels.

Like the JMP methodology, the ONEWASH Plus sustainability check framework combines different sub-indicators into a composite indicator. The highest service level in the JMP framework is 'Safely

Table 5. Service provider and authority indicator scores for ONEWASH Plus Programme towns.

Service provider indicator	Adishihu	Sheno	Maksegnit	Abomsa	Welenchiti	Kebridehar	Wukro	Average	Proportion of towns which score at least 50 (meeting the benchmark)
SP-I-1: Utility organisation	25	50	50	50	50	50	50	46	86%
SP-I-2: *Staff productivity*	50	75	0	50	50	0	50	39	71%
SP-I-3: Town Water Utility staffing	0	0	0	25	25	0	25	11	0%
SP-T-1: Non-revenue water	50	100	25	75	75	0	0	46	57%
SP-T-2: Adequate supply of spare parts for minor maintenance (pipes, fittings etc.)	75	50	25	25	50	25	75	46	57%
SP-T-3: Effective maintenance system in place	25	50	50	50	75	25	50	46	71%
SP-T-4: Water quality management and disinfestation	75	0	25	0	0	25	100	32	29%
SP-F-1: Cost recovery	50	50	50	50	50	0	50	43	87%
SP-F-2: Effective financial management	75	25	50	25	50	0	100	46	57%
SP-F-3: Effective asset management	0	0	25	0	0	0	0	4	0%
SP-F-4: Effective billing and collection	50	25	50	50	75	75	100	61	86%
SP-S-1: Urban poor get affordable water	25	25	25	25	25	25	25	25	0%
Average SP score	42	38	31	35	44	19	58		
Proportion of benchmarks met	58%	50%	42%	50%	67%	17%	67%		

Source: Authors.

Table 6. Service authority level indicator scores.

Service authority indicator	Adishihu	Sheno	Maksegnit	Abomsa	Welenchiti	Kebridehar	Wukro	Total	Number of towns which score at least 50
SA-I-1: Sufficient capacity at regional and zonal level to provide support to TWUs	75	50	50	50	50	25	75	54	86%
SA-I-2: Presence of Water Board (WB)	50	25	50	100	25	25	75	50	57%
SA-T-1: Effective provision of technical support to the TWU	50	50	50	50	50	25	50	46	86%
SA-T-2: Checks on construction quality	75	50	50	50	50	50	75	57	100%
SA-E-1: Catchment management system in place	0	0	0	0	0	0	0	0	0
Average SA score	50	38	38	38	38	25	50	41	
Proportion of benchmarks met	80%	60%	80%	80%	60%	20%	80%		

Source: Authors.

managed services', which implies access to improved water services on premise (accessibility), available when needed (reliability) and free of contamination (WHO, 2017). The sustainability check framework combines similar indicators, in addition to water quantity. Both use household survey

Table 7. National level indicator scores.

Indicator	Score and micro-scenario
N-I-1: National monitoring system or database	25: National database is in place, including asset and functionality data, but not updated annually
N-I-2: National support to regional, zonal and woreda/ service authority, including capacity building training and technical backup	25: There is ad-hoc/project based training of regional, zonal and woreda staff
N-I-3: Urban regulatory agency	0: No regulatory agency
N-T-1: National/regional standard/guideline/norms for urban water service provision	50: There are national or regional level norms and standards related to urban water service provision and they are known at regional, zonal, woreda level
N-F-1: Availability of national budget related to urban water supply	25: Less than 50% of required funds available
N-E-1: National environmental protection standards are established and applied to town water services	25: National standards to protect environment in design, sizing and siting of water supply infrastructure, but not known and/or enforced

Source: Authors.

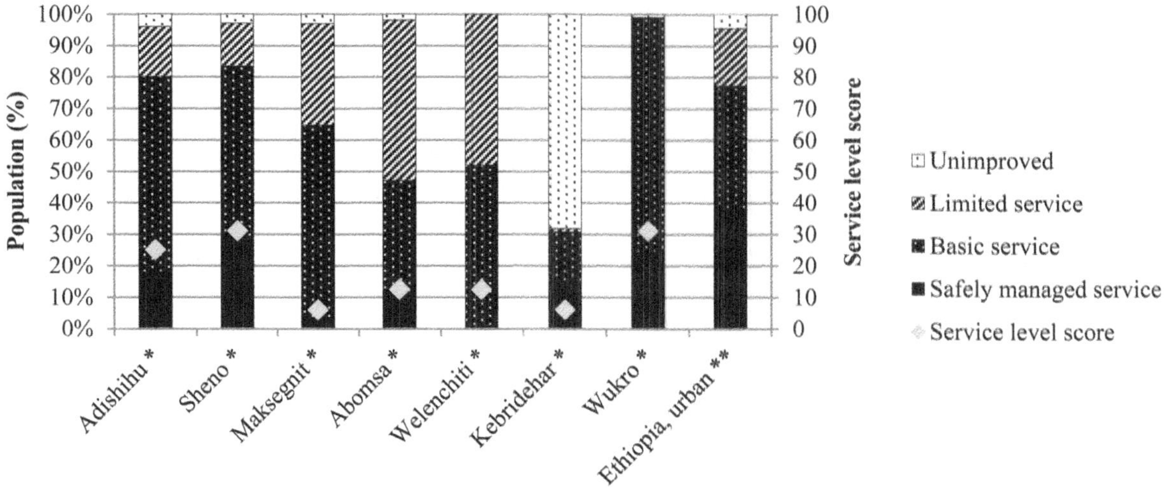

Fig. 2. Water service levels. *Source:* Compiled by authors based on (*) ONEWASH Plus 2014–2015 household survey and sustainability check data and (**) 2015 JMP data (WHO/UNICEF, 2017).

data as well as data from service providers or regulatory bodies. However, where the emphasis of JMP is on household survey data, the sustainability check mainly depends on data obtainable from service providers.

As shown in Figure 2, application of the JMP framework and the sustainability check framework shows similar trends in service levels, with highest levels in Wukro, Sheno and Adishihu and lowest in Maksegnit. The fact that Welenchiti scores higher under the sustainability check framework than under the JMP framework is due to the fact that water quantity, on which the town scores relatively well, was included in the sustainability framework and not in the JMP framework.

5.2. Main small town challenges

The low scores on various sustainability indicators reflect some of the main issues that small towns face in water service provision. Looking 'horizontally' at Tables 5 and 6, the towns score lowest on the following indicators: Town water utility staffing; Effective asset management; Initiatives for the urban poor; Water quality management and disinfection; Catchment management.

Small town staffing and capacity. One of the major challenges that these towns face is in attracting, employing and training operational and managerial staff (Adank, 2013). The ONEWASH Plus sustainability check showed low scores on the 'staffing' indicator, with four of the seven towns having an understaffed Town Water Utility with less than 75% of the required staff in place. Capacity of the existing staff was also observed as a potential sustainability challenge. In none of the towns had all staff members received training. Only in Adishihu, Maksegnit and Abomsa had at least half of the operational staff received training, while half of the administrative staff had been trained in Adishihu, Welenchiti and Wukro. In Abomsa, only financial staff had received training.

Asset management. A good overview of assets and their state is crucial for sustainable water service provision. This facilitates planning of timely repairs and rehabilitation of scheme components and is especially important for more complex urban and small town schemes. Asset management has proven a big challenge and is a potential sustainability risk in the assessed small towns. This is partly due to challenges with the handing over process of the infrastructure. Construction supervisors, usually consultants under contract with the Regional Water Bureau, are supposed to provide construction completion reports, which include an overview of all developed assets. However, these construction completion reports are often not submitted, or are only submitted to the Region and not included in the handover of the infrastructure to the town and are therefore often not available to Town Water Utilities. Another challenge is the fact that it is not always clear who is the asset holder and therefore responsible for asset renewal and replacement. As the water policy states that water infrastructure is owned by the state, the town councils representing the public in the service areas can be considered the asset holders. However, Town Water Utilities are also expected to contribute to asset expansion and renewal. Rules and arrangements in this area have yet not been clearly spelled out.

Initiatives for the urban poor. Water service provision to the poor is often not high on the agenda of small town water service providers, and this was also the case for the seven Town Water Utilities discussed in this paper. All provided water services through public taps, which are considered as pro-poor solutions. However, with the exception of Wukro, none of the utilities had made provisions for shared yard connections, which allow the urban poor living in compound housing to jointly apply and pay for a connection. Although shared connections were allowed in Wukro, the number of public taps was very limited.

Water quality management and disinfection. Water quality management practices were found to differ between the regions. The three Oromia towns did not practise disinfection of the ground or elevated storage reservoirs. In the Amhara town, Maksegnit, and the Somali town, Kebridehar, disinfection of storage reservoirs was practised, but less often than once a month. In the Tigrayan towns Wukro and Adishihu, monthly disinfection of reservoirs was undertaken by a qualified operator and water quality (chemical, bacteriological, physical) was checked on irregular (less than once a month) basis.

Catchment management. Growing small towns put increasing demands on water resources. Catchment management is important to protect quality and quantity of the water source, as well to avoid conflicts between different water users. Not having any arrangements in place, as was found to be the case in the study towns, could pose a serious environmental sustainability risk.

Enabling environment at national level. Small town water supply often falls between the cracks between urban and rural when it comes to setting clear policies and laws. This is often the case in contexts where one national ministry, department or agency is dedicated to rural and one to urban water supply, with the small town sub-sector falling in either or neither (Adank, 2013). Various authors consider regulation of small town water services a major weakness (Adank, 2013; Pilgrim *et al.*, 2007; WSP, 2010; Pezon, *et al.*, 2013) as regulatory arrangements for small town water services are often missing, unclear or, in case of multiple service delivery models in the same town, in contradiction with each other. In small town schemes, the asset owner (generally local government) often acts as the local regulatory body for those aspects most directly related to service provision, such as tariff setting (Pilgrim *et al.*, 2007) and performance regulation. This is also the case in Ethiopia where the absence of a regulatory agency was identified as an urban and small town weakness.

5.3. Use and conditions for use of sustainability checks

The sustainability check developed and applied under ONEWASH Plus is a project-based tool, and has been used for monitoring progress on project outcomes. In addition it has informed simple sustainability plans, with actions for mitigating identified sustainability challenges. Scoring of the indicators has so far been done by project staff based on data they collected from Town Water Utilities and staff from the Regional Water Bureaux. In subsequent annual rounds, the project intends to involve these stakeholders further in data collection and indicator scoring processes. This is intended to provide staff with deeper insight into challenges that need to be addressed.

In the absence of clear regulatory mechanisms, sustainability check results can provide Regional Water Bureaux with a tool for regulating the performance of Town Water Utilities and the services that they provide. Principles, indicators and methodologies developed as part of the sustainability check could be incorporated into and linked to regional and/or national monitoring systems and regulatory arrangements, which are currently under development in Ethiopia.

The dissemination and sharing of the results of the checks at national level can inform and influence policy discussion and change. The presentation of the first sustainability check at the 7th Multi Stakeholder Forum, organised by the One WASH National Programme has for example contributed to the discussion on the need for an urban and small town regulatory framework.

The costs of the first round of sustainability checks, which, in addition to small town water services included institutional WASH services, rural water and sanitation services in the villages around the towns, and town sanitation services, amounted to about 14,000 USD per town (about 0.72 USD per town inhabitant). This is equated to about 1.2% of the total project implementation costs. This is within the 0.8% to 2.8% range reported as the costs of a range of sustainability checks (Schweitzer, 2016). Sustainability checks are considered to be a cost effective means for measuring and influencing the sustainability of water services in small towns. Inclusion of sustainability check indicators into the national and regional monitoring, audit and reporting frameworks could provide substantial economies of scale and greater impact. Uptake by Regional Water Bureaux of roles in data compilation, processing and analysis at scale is estimated to cost about 42,000 USD per year per region (about 430 USD per town). This is considered good value, but to put this figure into perspective, current operational budgets of Regional Water Bureaux are only about 85,000–125,000 USD. Application of sustainability checks at scale through government structures will likely only be achieved when more financial (and human) resources are allocated to recurrent expenditure and support activities, as opposed to construction of new infrastructure.

6. Conclusions

This paper has presented a framework for assessing and monitoring small town water services provision. It combines elements from rural sustainability checks and urban benchmarking frameworks and methodologies. Application in seven towns has shown the usefulness of the framework and generated actionable insights for small and medium towns.

Use of the sustainability check framework has revealed discrepancies between the ideal and the current situation, both in provided levels of service, and in terms of conditions for sustainable water service provision at service provider, service authority and national levels. While a useful tool for identifying service provision and sustainability challenges, whether the capacity and motivation are in place to address these challenges beyond the scope of the ONEWASH Plus Programme remains to be seen.

At the moment, execution of the sustainability checks as applied in the ONEWASH Plus project context is human resource and capital intensive and therefore relatively difficult for service providers (Town Water Utilities) and service authorities (Water Boards, Town Administrations, Regional Water Bureaux) to take up themselves. Elements of sustainability checks, including indicators and parameters related to service levels, performance of service providers and service authorities, should ideally be integrated into national and regional M&E and reporting frameworks. If such indicators and parameters were part of routine, continuous monitoring and reporting, the costs of application of the sustainability checks could be reduced considerably. When applied at scale, sustainability checks would also have the potential to inform regulation. This could involve use of sustainability checks by a national or regional regulator, assessing water service provision against norms and standards and enforcing these. Dissemination of sustainability check results, especially when done at scale also has the potential to assist water service users in holding their service providers and authorities accountable for their performance and the services that they provide.

References

Adank, M. (2013). *Small Town Water Services: Trends, Challenges and Models.* Thematic Overview Paper No. 27. IRC, The Hague, The Netherlands.

Adank, M., Butterworth, J., Godfrey, S. & Abera, M. (2016). Looking beyond headline indicators: water and sanitation services in small towns in Ethiopia. *Journal of Water Sanitation and Hygiene for Development 6*(3), 435–446. doi:10.2166/washdev.2016.034.

Bain, R., Gundry, S., Wright, J., Yang, H., Pedley, S. & Bartram, J. (2012). Accounting for water quality in monitoring access to safe drinking-water as part of the Millennium Development Goals: Lessons from Five Countries. *Bulletin of the World Health Organization 90*(3), 228–235.

Bartram, J. (2008). Improving on haves and have-nots. *Nature 452*, 283–284.

Boulenouar, J. (2016). *Rapid Scan of Nine Years of Sustainability Checks. Lessons Learned and Next Steps.* AguaConsult, Wivenhoe, UK.

Boulenouar, J., Schweitzer, R. & Lockwood, H. (2013). *Mapping Sustainability Assessment Tools to Support Sustainable Water and Sanitation Service Delivery.* Triple-S Project Working Paper No. 6. IRC, The Hague, The Netherlands. Available at: http://www.ircwash.org/resources/mapping-sustainability-assessment-tools-support-sustainable-water-and-sanitation-service (accessed 21 February 2017).

Central Statistical Agency (CSA) (2013). *Population Projection of Ethiopia for All Regions at Woreda Level from 2014–2017.* CSA, Addis Ababa, Ethiopia.

Dutch WASH Alliance (undated). Measuring the likelihood of WASH sustainability. http://wash-alliance.org/wp-content/uploads/sites/36/2017/03/Sustainability-Monitoring-Framework-2-pager.pdf (accessed 27 September 2017).

Federal Democratic Republic of Ethiopia (2013). *One WaSH National Program, A Multi-Sectoral SWAp, Programme Document.* Federal Democratic Republic of Ethiopia, Addis Ababa, Ethiopia.

Federal Democratic Republic of Ethiopia (2015). *(Unpublished Draft) Water Section for Second Growth and Transformation Plan (GTP II, 2016–20).* Federal Democratic Republic of Ethiopia, Addis Ababa, Ethiopia.

Godfrey, S. & Labhestwar, P. (2010). How safe are the global water coverage figures – a case study from Madhya Pradesh, India. *Environmental Monitoring and Assessment Journal 176*, 561–574.

Godfrey, S., Van der Velden, M., Muianga, M., Xavier, A., Downs, K., Morgan, C. & Bartram, J. (2015). Sustainability check: five-year annual sustainability audits of the water supply and open defecation free status in the 'one million initiative', Mozambique. *Journal of Water Sanitation and Hygiene for Development 4*(3), 471–483. doi: 10.2166/washdev.2014.118.

Independent Commission for Aid Impact (ICAI) (2016). *Report: Assessing DFID's Results in Water, Sanitation and Hygiene. An Impact Review.* Online. Available from: http://icai.independent.gov.uk/html-report/wash/ (accessed 15 September 2016).

Kayser, G. L., Moriarty, P., Fonseca, C. & Bartram, J. (2013). Domestic water service delivery indicators and frameworks to monitoring, evaluation, policy and planning: a review. *International Journal of Environmental Research and Public Health 10*(10), 4812–4835.

Lockwood, H. & Smits, S. (2011). *Supporting Rural Water Supply: Moving Towards A Service Delivery Approach.* Practical Action Publishing, Warwickshire, UK.

Ministry of Urban Development and Housing (MUDH) (2013). *Towns Classification and Categorization Directive.* Ministry of Urban Development and Housing. Addis Ababa, Ethiopia.

Ministry of Water Resources, Irrigation and Electricity (MOWIE) (2013). *National Guidelines for Urban Water Supply and Sewerage Services.* Ministry of Water, Irrigation and Electricity, Addis Ababa, Ethiopia.

Ministry of Water Resources, Irrigation and Electricity (MOWIE) (2014). *(Unpublished Draft) Criteria for Allocation of Funds for Urban Water Supply Investment.* Ministry of Water, Irrigation and Electricity, Addis Ababa, Ethiopia.

Moriarty, P., Patricot, G., Bastemeijer, T., Smet, J. & van der Voorden, C. (2002). *Between Rural and Urban: Towards Sustainable Management of Water Supply Systems in Small Towns in Africa.* Working Paper. IRC International Water and Sanitation Centre, Delft, The Netherlands.

Murungi, C. & Blokland, M. W. (2016). Benchmarking for the provision of water supply and sanitation services to the urban poor: an assessment framework. *International Journal of Water 10*, 155. doi:10.1504/IJW.2016.075566.

Onda, K., LoBuglio, J. & Bartram, J. (2012). Global access to safe water: accounting for water quality and the resulting impact on MDG progress. *International Journal of Environmental Research and Public Health 9*, 880–894.

Pelto, P. & Pelto, G. (1978). *Anthropological Research: The Structure of Inquiry.* Cambridge University Press, Cambridge and New York.

Pezon, C., Agognon, F. & Bassono, R. (2013). *Coût Performance et Régulation des Petites Réseaux de Distribution D'eau Potable au Burkina Faso.* Triple-S Document de Recherche 2. IRC, The Hague, The Netherlands.

Pilgrim, N., Roche, B., Kalbermatten, J., Revels, C. & Kariuki, M. (2007). *Principles of Town Water Supply and Sanitation, Part 1: Water Supply.* Water Working Note 13, The World Bank, Washington, DC, USA.

Rijksdienst voor Ondernemend Nederland RVO.nl (2016). Annex 3 to administrative rules FDW, 3rd call: Sustainability framework (FIETS kader). https://english.rvo.nl/sites/default/files/2016/08/FDW16%20beleidsregels%20Annex%203%20FIETS%20kader.pdf.

Ssozi, D. & Danert, K. (2012). *National Monitoring of Rural Water Supplies: How the Government of Uganda Did it and Lessons for Other Countries*, RWSN-IFAD Series 4. Rural Water Supply Network, St Gallen, Switzerland.

Schweitzer, R. (2016). *Case Study: Sustainability Index Tool.* Presentation presented at Learning Seminar: Monitoring for Sustainability, 9 August 2016, Addis Ababa.

USAID (2013). *WASH Sustainability Index Tool.* http://washplus.org/rotary-usaid.html (accessed 27 September 2017).

Van den Berg, C. & Danilenko, A. (2010). *The IBNET Water Supply and Sanitation Performance Blue Book: The International Benchmarking Network for Water and Sanitation Utilities Databook.* World Bank, Washington, DC.

Water and Sanitation Program of the World Bank (WSP) (2010). *Sustainable Management of Small Water Supply Systems in Africa Practitioners' Workshop Report. Field Note.* WSP, Nairobi, Kenya. Available at: http://www.wsp.org/sites/wsp.org/files/publications/Sustainable_Management_of_small_water_supply_ systems_in_Africa_English.pdf (accessed 17 July 2013)

WHO/UNICEF Joint Water Supply and Sanitation Monitoring Programme (2014). *Progress on Drinking Water and Sanitation – 2014 Update.* World Health Organization, Geneva.

WHO/UNICEF Joint Water Supply and Sanitation Monitoring Programme (2017). *Progress on Drinking Water, Sanitation and Hygiene: 2017 Update and SDG Baselines.* World Health Organization, Geneva.

van Wijk-Sijbesma, C. (2001). *The Best of Two Worlds? Methodology for Participatory Assessment of Community Water Services.* Wageningen Agricultural University/IRC, Wageningen/ The Hague, The Netherlands.

World Bank (2015). *Ethiopia Urbanization Review: Urban Institutions for A Middle-Income Ethiopia.* World Bank, Washington, USA.

World Health Organization (WHO) (2017). *Safely Managed Drinking Water – Thematic Report on Drinking Water.* World Health Organization, Geneva, Switzerland.

In the shadow of the city: financing water infrastructure in small towns in Burkina Faso

Elena Humphreys[a,*] and Klaas Schwartz[a,b]

[a]*IHE Delft Institute for Water Education, Westvest 7, 2611AX, Delft, The Netherlands*
**Corresponding author. E-mail: elenahumphreys@gmail.com*
[b]*Department of Human Geography, Planning and International Development Studies, Amsterdam Institute for Social Science Research, University of Amsterdam, Nieuwe Achtergracht 166, Amsterdam 1018 WV, The Netherlands*

Abstract

The abundant praise awarded for the development of the urban water services sector in Burkina Faso stands in stark contrast with the development of the rural water services sector. This article examines the funding of water infrastructure in four small villages in Burkina Faso. The article finds that public funding for water infrastructure for these municipalities is largely nonexistent. First of all, central government makes very little funding available for rural areas. Funding that is made available is then also prioritized for regions that already are relatively well covered. Secondly, the municipalities themselves also prioritize other sectors over the water sector for the investment of locally generated revenue. As a result, these municipalities rely on donor funding for developing water supply in their villages. This dependence not only leaves these municipalities vulnerable to shifts in donor funding but can also lead to inequalities as some municipalities are better at attracting donor funds than others. Some small towns are thus confronted with a double bias. First an urban bias in which the majority of public finance goes to urban centres. Second, by a donor-bias in which some towns are favoured for project implementation due to favourable site characteristics.

Keywords: Financing; Water infrastructure; Small towns; Burkina Faso

1. Introduction

In 2015, the Joint Monitoring Program estimated that 663 million people still lack access to improved water sources (WHO/UNICEF JMP, 2015)[1]. The same report highlights how Burkina Faso met the

[1] The term 'improved water sources' was developed to allow for international comparability for monitoring progress on the Millennium Development Goals (MDGs) by the World Health Organization (WHO) and United Nation's Children Fund (UNICEF) Joint Monitoring Program. Improved water sources include piped water (to dwelling or yard), standpipes, boreholes, dug wells, protected springs, rainwater, and bottle water (WHO/UNICEF JMP, n.d.).

doi: 10.2166/wp.2018.005

Millennium Development Goal (MDG) target on drinking water by suggesting that 61% of the population gained access since 1990 resulting in an overall water coverage figure of 82%. The discussion of the impressive water coverage achievements of Burkina Faso are usually accompanied by a discussion of ONEA, the *Office National de l'Eau et de l'Assainissement* (ONEA), which is responsible for water supply provisioning in 50 urban centres in Burkina Faso. ONEA is often commended on how, through performance-based management, the water utility has managed to improve performance. In fact, ONEA managed to increase water coverage for its areas from about 50% in 2001 to well over 80% by 2014. Particularly impressive are the coverage figures for the capital city Ouagadougou. Despite being Sub-Saharan Africa's fastest growing city with an annual growth rate of 7.5%, ONEA managed to increase service coverage from 50% in 2000 to 86% in 2014[2] (van den Berg & Danilenko, 2017). Moreover, ONEA has been able to do this whilst still being able to cover operational costs and part of its investment costs (van den Berg & Danilenko, 2017).

Impressive as the performance of ONEA is, the abundant praise of ONEA obscures a very different water supply reality that characterizes rural areas of Burkina Faso. Whereas urban water coverage exceeds 97%, rural water coverage reaches only 76% (WHO/UNICEF JMP, 2015). If national statistics are used instead of JMP figures, the coverage rate drops to 65 percent, which is even short of the 74 percent target established for the MDGs (INSD & MEA, 2015). Moreover, according to figures presented by the WHO/UNICEF Joint Monitoring Program the percentage of the rural population that have access to water drops to 0% when the criterion of piped on-premise connections is added. This means that gains in water coverage has mainly been through low-cost technologies. The urban-rural imbalances in water coverage are also reflected in the financial contributions from the national Government of Burkina Faso for water infrastructure development. The contribution of the central government to water services show a strong bias towards water supply in urban centres despite the fact that between 69% and 77% of the Burkinabe population lives in areas that are officially classified as being rural[3]. As highlighted by Figure 1, from 2008 to 2014 between 68% and 83% of central government expenditure on water was allocated to urban centres. The financing bias is also visible in the financing gaps for rural and urban water supply. A study into the financing of water services in Burkina Faso suggested that the financing gap for rural water supply stood at US$ 48 million, whilst the financing gap for urban water supply amounted to US$ 28 million (WSP, 2011:9). Given that per capita cost of service provision in rural areas are higher than in urban areas due to lower economies of density and economies of scale (Pietila, 2006) the financing bias translates into even greater inequalities in terms of the levels of service enjoyed by consumers.

In this article, we focus on this neglected side of the Burkina Faso water sector by examining the funding of water infrastructure in small villages in Burkina Faso. This is done by examining water infrastructure funding of four municipalities: Houndé, Komsilga, Banfora and Moussoudougou. Developing these case studies was done by undertaking a total of 43 semi-structured interviews with local, regional, and central governmental officials and (inter)national water sector experts between November 2016 and February 2017.

---

[2] In 2000, ONEA served half of 2.3 million people. In 2014 ONEA served 4 million people out of a total population of 4.6 million people (van den Berg & Danilenko, 2017:109).

[3] https://data.worldbank.org/indicator/SP.RUR.TOTL.ZS?locations = BF&view = chart (accessed 30 August, 2017) and INSD (2016).

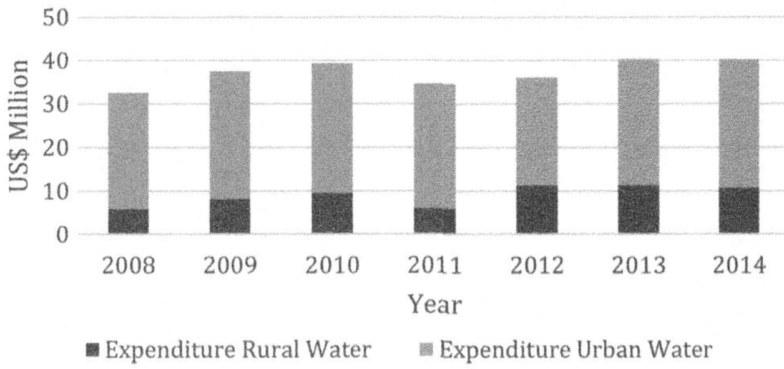

Fig. 1. Central government expenditure on urban and rural water in 2008–2014. *Sources:* MEA, 2016; OANDA, 2016; World Bank, 2016.

The article finds that public funding for water infrastructure for these municipalities is largely non-existent. First of all, central government makes very little funding available for rural areas. Funding that is made available is then also prioritized for regions that already are relatively well covered. Secondly, the municipalities themselves also opt to not invest locally generated revenue into water supply infrastructure. Instead other sectors, such as health and education, are prioritized by municipalities. Instead of public funding, these municipalities chose to rely heavily on external aid and donor funding for developing water supply in their villages. This dependence has two important consequences. First, it leaves these municipalities vulnerable to shifts in donor funding and requires municipalities to accept donor objectives and demands in order to access these funds. Secondly, not all municipalities have the same ability to attract external aid funding. Particularly the least developed municipalities appear to have the greatest difficulty in attracting sufficient funding for the development of water infrastructure. This can contribute to growing inequities regarding water services between municipalities in Burkina Faso.

In developing this article, we first elaborate on the institutional arrangements in the Burkinabe water sector. In particular we explain the public financing structure of the water sector. Next the article discusses the four case studies and identifies the different sources of funding that these towns have access to. The article continues by analyzing what factors allow a municipality to be an attractive partner for donors. The article concludes with a discussion of these findings and the consequences that this entails for the four towns in terms of the development of their water systems.

2. Institutional arrangements and public financing of water supply in Burkina Faso

The responsibilities for water and sanitation service provision are shared among a number of organizations. Below the different organizations and with their responsibilities in terms of water and sanitation service provision are presented and summarized in Figure 2. In discussing these arrangements, particular focus is on the role of these organizations in the public funding of water services.

2.1. Ministry of Water and Sanitation and the General Directorate for Potable Water

The Ministry of Water and Sanitation (MEA) is the national government body responsible for the development of the national priorities for water and sanitation in Burkina Faso. Within the Ministry,

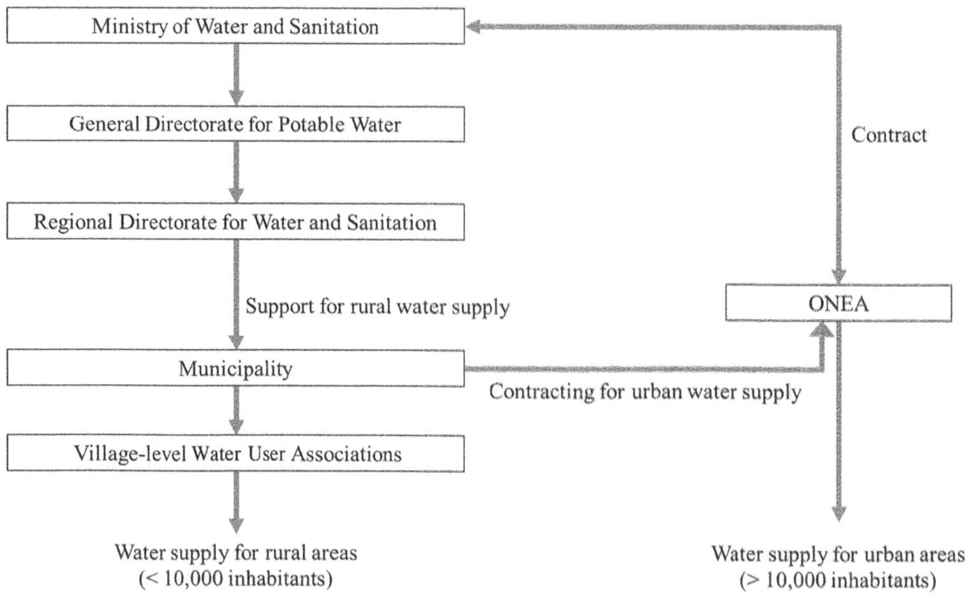

Fig. 2. Institutional arrangement for water supply in Burkina Faso. *Source:* primary data.

the General Directorate for Potable Water (DGEP) is responsible for water supply[4]. To achieve the MDGs, the MEA and DGEP[5] established the 2007–2015 National Program for Water and Sanitation (DGRE, 2006). In light of the SDGs, the DGEP has established the 2016–2030 National Program for Potable Water focusing specifically on the provisioning of potable water (DGEP, 2015). The 2016–2030 National Program for Potable Water has an estimated cost of about US$ 2 billion and aims to achieve universal access to potable water by 2030. The DGEP monitors progress in the sector, searches for funding, and maintains the requirements of funding agreements[6].

The Ministry of Water and Sanitation has two ways of channelling financial resources to municipalities. One channel transfers funds to the Regional Directorate for Water and Sanitation (DREA), which then distributes the funds to municipalities. The second channel directly funds the municipality for water infrastructure development. Between 2007 and 2015, the national government transferred just a small proportion of their public expenditure to the DREAs or directly to the municipalities. The national government transferred 20 percent of public expenditure for the DREA. However, the transfers to DREAs also support the operations and administration of these organizations, which means that fewer than the 20 percent resources transferred were actually available for water infrastructure development at the municipal level[7]. Direct transfers for the development of water infrastructure to municipalities only represented 8 percent of public expenditure on water supply to municipalities.

---

[4] Interview DGEP, December, 2016.

[5] At the time, the responsibilities of the Ministry of Water and Sanitation and the General Directorate for Potable Water were under the Ministry of Agriculture, Water and Hydraulic Resources and the General Directorate for Water Resources (Interview USAID, December, 2016).

[6] Interview DGEP, December, 2016.

[7] Interview CDREA2, December, 2016.

2.2. Regional Directorate for Water and Sanitation

Some national functions related to the 2007–2015 National Program for Water and Sanitation were deconcentrated to the DREAs (DGRE, 2006). Each of the 13 regions have a DREA, which supports the municipalities in their water and sanitation service provision. Specifically in the 2007–2015 National Program for Water and Sanitation, the DREAs are responsible for: (1) coordinating implementation; (2) supporting the municipalities in developing their municipal development plans for water and sanitation; (3) measuring progress towards development objectives; and (4) distributing financing for infrastructure development (DGRE, 2006; MEA & DGESS, 2016). In the new 2016–2030 National Program for Potable Water, DREAs are responsible for constructing new water tower systems and multi-village water distribution systems, while continuing their support for municipalities and promoting private-public partnerships (DGEP, 2015).

2.3. National Office for Water and Sanitation

ONEA is a state-run company that is responsible for water and sanitation service provision in urban areas, which is defined as settlements of 10,000 people or more (ONEA, 2016). The provision of urban water services by ONEA can be organized in two ways. First, ONEA can provide direct water service provision through a three year contract with the Ministry of Water and Sanitation[8]. There are currently 50 urban centres serviced by ONEA under this contract with the National Government[9]. Second, municipalities may delegate water service provisioning to ONEA through a contract that is entered into by ONEA and the municipality. This contract is managed at the municipal level. There are currently eight municipalities that have a lease agreement with ONEA[10]. ONEA has two basic service models: through private connections and by way of standpipes that are serviced through water towers[11]. The water is sold volumetrically and tariffs are established by the National Government[12]. However, in municipalities where ONEA is delegated water service provisioning through a municipal-level contract, fees can be added to the tariffs if the municipality chooses to do so[13]. User tariffs support the maintenance and operations of water infrastructure of ONEA[14]. In addition, these tariffs support investment cost, which are typically made using matching funds from the government or partners[15]. Decisions about extending services are highly centralized, requiring approval from the central office of ONEA in Ouagadougou before funding is mobilized[16]. In municipalities that have leasing contracts with ONEA, the municipalities can decide to extend services, but the financing used to expand services is the responsibility of the municipality (Commune de Hounde & GIZ, 2009)[17].

---

[8] Interview ONEA, December, 2016.
[9] Interview ONEA, December, 2016.
[10] Interview ONEA, December, 2016.
[11] Interview ONEA, December, 2016.
[12] Interview ONEA, December, 2016; DGEP, December, 2016.
[13] Interview ONEAH, December, 2016.
[14] Interview ONEA, December, 2016.
[15] Interview IRCD, December, 2016.
[16] Interview ONEAB, November, 2016.
[17] Interview ONEAH, December, 2016.

2.4. Municipalities and Water Users Associations

Burkina Faso is divided into 13 administrative regions, which are comprised of 45 provinces, made up of 351 municipalities known in Burkina Faso as 'communes' (INSD, 2016)[18]. Each municipality is comprised of a number of villages, while each village is comprised of neighbourhoods. Municipalities are responsible for water service provision in the villages in their municipality. Each municipality should have a plan for water infrastructure development. The actual operation and management of water infrastructure is delegated to Water Users Associations, which are installed at the village level (MAHRH, 2006). Water Users Associations are delegated responsibility by the municipality for the regular maintenance and operations of the water services within their village[19]. The regular maintenance and operation is financed by monthly service fees collected by the Water Users Association[20], which are used to fund preventative maintenance and provide 'good faith' payments when funding is transferred for large repairs (US$ 281) or new infrastructure (US$ 441)[21].

3. Case studies: finance for water infrastructure development

'Why do you want the budget? There is nothing in there for water.'
Municipal Accountant, Banfora

The municipalities studied all have both urban and rural water infrastructure characteristics in the sense that the municipalities have small urban centers which are serviced by a small network, private connections and standpipes, as well as less densely populated areas serviced by standpipes (overview in Table 1). The municipalities incorporate between four (Moussodougou) and 36 (Komsilga) villages. Water coverage for the non-urban population of these municipalities varies from 63.8% to 100%. Banfora and Houndé are the only municipalities in which ONEA is active. As ONEA needs to cover its operational costs from tariffs, it requires a sufficient market-size, which is only available in Houndé and Banfora. These municipalities have entered into a contract with ONEA to serve the urban parts of the municipality. For villages not covered by ONEA the Banfora and Houndé have engaged Water Users Associations.

3.1. National government expenditure in case study towns

The national government is responsible for financing the development of rural water infrastructure by transferring resources to the municipalities or to the regions. As highlighted earlier, however,

[18] Municipalities are called 'communes' in Burkina Faso. The terms are interchangeable so any reference to 'communes' or 'commune' in this article refers to municipalities or municipality.
[19] BST1, November, 2016.
[20] There is a basic fees structure suggested by the Ministry of Water and Sanitation, however each Water Users Association is free to alter the structure, given that sufficient revenue is generated to meet 3000 FCFA per household. Each Water Users Associations visited had a different arrangement for fee collection (e.g. based on the amount of spouses, through the mutualization of profits from agricultural group, or direct payments from the sale of cotton) (Interview BKAUE1, November, 2016; Interview HBAUE, December, 2016).
[21] Interview BST1, November, 2016.

Table 1. Overview of case studies.

Indicator	Houndé	Komsilga	Banfora	Moussodougou
Population	114,774	65,432	162,055	15,746
Urban (ONEA)	67,232		110,891	
Rural	47,542		51,164	
Water coverage rural population	63.8%	99.4%	75.4%	100%
Villages	15	36	22	4
Wells	233	404	312	58
Functionality of wells	92.9%	87.3%	87.2%	100%
Water towers	6	13	3	3
Functionality of water towers	100%	94.2%	100%	100%

Sources: Commune de Moussodougou, IGIP Afrique & Oxfam Quebec, 2013; INSD & MEA, 2015; Commune de Banfora & Oxfam Quebec, 2013; Commune de Hounde & GIZ, 2009; Commune de Komsilga & GRAD Consulting Group, 2013.

transfers from the national government to non-urban areas are limited. Of the four cases studies, only Houndé received a transfer from the Ministry of Water and Sanitation of US$ 58,979 for five new wells (see Figure 3). This represents approximately 21% of the total costs for planned water infrastructure in the municipality. The other municipalities received no funding from the national government for water infrastructure development. The challenges of funding water infrastructure development are illustrated by the municipal Secretary General of Banfora. Referring to the decentralization of responsibilities for water provision to the municipal level, he observed that the national government '*transferred the problems, not the resources*'[22]. This perspective was countered by a representative of the General Directorate of Potable Water. He argued that it was necessary for a municipality to have administrative capacity before public funds could be transferred to that municipality[23]. The urban bias in public finance is then linked to the capacity of the municipality to absorb the funds from the central government.

3.2. Municipal expenditure on water infrastructure

With national government expenditure in the case study sites limited to a small contribution for Houndé, the role of the municipality in funding development of water infrastructure becomes more important. However, only the municipality of Komsilga provided a funding contribution for the construction of water infrastructure. This contribution amounted to US$ 1,160, which represents a mere 0.14% of the cost of the water infrastructure plan. The other municipalities did not allocate any budget for water infrastructure development (see Table 2). Two factors explain the limited budget allocation of the municipalities.

The first factor relates to the challenges municipalities face in local revenue generation. The municipalities are characterized by low capacity and motivation to administer taxes by municipal councillors, difficulties of collection of taxes, and the municipal tax authorities favour urban settings.

[22] All quotes translated by the researcher.

[23] Interview DGEP, December, 2016; Interview KAM, December, 2016.

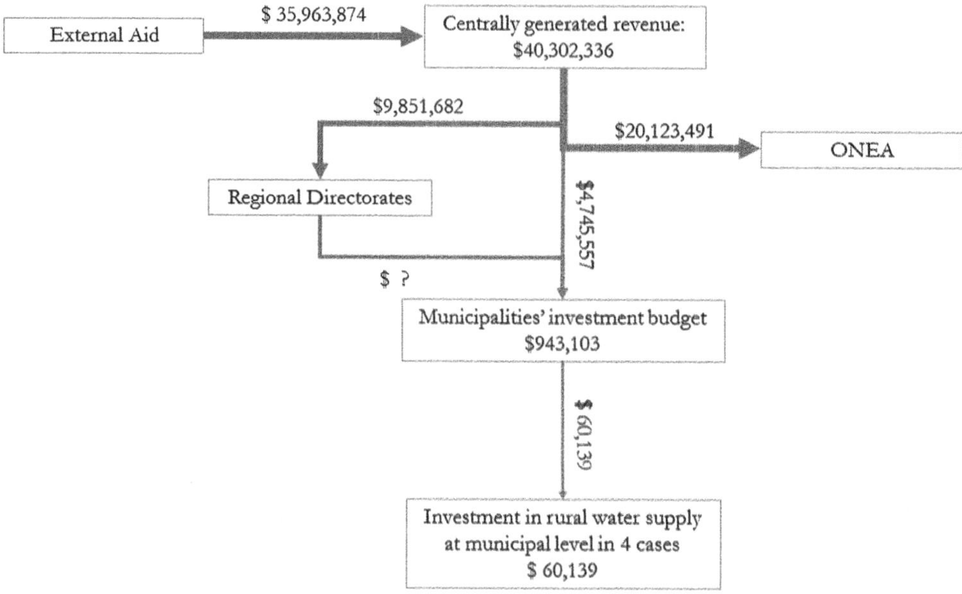

Fig. 3. Public funding of water infrastructure in case study towns.

These characteristics mean that the local revenue generation is low (see also Humphreys *et al.*, 2018). A comparison of the locally generated revenue to the annual cost of the municipal development plans' water infrastructure development, reveals that local revenue generation is insufficient to make annual progress on each municipality's plan for water infrastructure development. The proportion of total municipal budget needed for water infrastructure development varies from over 125 percent in Banfora to 17 percent in Komsilga. Komsilga has the highest local revenue generation out of the case study municipalities and therefore, would be best situated to use locally generated revenue for water. Komsilga's proximity to the capital of Ouagadougou increases their local revenue generation capacity as the price of land is higher in Komsilga (Commune de Komsilga & GRAD Consulting Group, 2013). The price of land affects the revenue base for municipal taxes, such as on land transfers. However, Komsilga only allocates less than 1% of the investment needed to develop planned water infrastructure from its municipal budget. To make annual progress on the municipal development plan, Komsilga would have to allocate nearly one-fifth of their total municipal budget to water infrastructure development.

This introduces the second factor that explains the limited allocation of municipal resources. In addition to the weakness of local revenue generation, municipalities prioritize other sectors for the use of municipal resources over the water services sector. From this analysis, it is clear that the resource generation at the municipal level is minimal and is unable to support the development of water infrastructure. The municipalities prioritize other sectors, such as health, education and agriculture for the allocation of municipal resources (WaterAid, 2014:10). Although municipalities are responsible for ensuring water services provision in, they do not view investment in water infrastructure as their responsibility. Rather, investment in water infrastructure is seen as the responsibility of donors.

Table 2. Public funding and the municipal development plan (in US$).

	Houndé	Komsilga	Banfora	Moussodougou
Total revenue per capita	4.57	17.76	1.56	7.07
Operations budget	306,563.16	617,595.62	166,582.87	17,353.12
Investment budget	218,509.34	544,706.85	85,838.36	94,048.30
Cost of water infrastructure in plan	1,391,632.89	802,977.76	1,611,731.09	No cost estimate included
Cost of water infrastructure in plan per year	278,326.58	200,744.44	322,346.22	(see above)
% of investment budget needed	127.38%	36.85%	375.53%	(see above)
% of total revenue needed	53.01%	17.27%	127.70%	(see above)
Allocations from national government	58,979.46	0	0	0
Allocations from local resources	0	1,160	0	0

Sources: Commune de Banfora, 2016; Commune de Banfora & Oxfam Quebec, 2013; Commune de Hounde, 2016; Commune de Hounde & GIZ, 2009; Commune de Komsilga, 2016; Commune de Komsilga & GRAD Consulting Group, 2013; Commune de Moussodougou *et al.*, 2016a; Commune de Moussodougou, 2016b; INSD & MEA, 2015; OANDA, 2016; World Bank, 2016.

4. Donor funding: partner finance and project finance

'That is a job for the partners.'
Mayor of Houndé

The combination of low local revenue generation and minimal local allocation for water, and the limited transfers from the national government mean that other sources of funding have to be found by the municipalities to develop water infrastructure in their villages. These other sources of funding for water infrastructure come in two forms: partner finance and project finance (see Table 3). Partner finance and project finance require varying levels of involvement from the municipalities and create accountabilities to stakeholders outside of the municipalities.

4.1. Partnership finance

Partners essentially concern development agencies or 'northern' municipalities that have entered into a partnership with the Burkinabe municipalities to support the development of the municipality. As such, partner finance is oriented to assisting the municipality in developing capacity over the length of their partnership[24]. Partners' finance activities are usually in accordance with the municipal development plans; typically because the partners have helped the municipality develop the municipal development plan (Commune de Hounde & GIZ, 2009; Commune de Banfora & Oxfam Quebec, 2013; Commune de Moussodougou *et al.*, 2016a)[25]. Accordingly, the financial resources used in this type of partnership are kept in an account that is jointly managed by the municipality and the partner[26]. While these partnerships are aimed to develop autonomous managerial capacity at the municipality, the service technicians in Houndé and Moussodougou expressed concern about the forthcoming end of their

[24] Interview HST, December, 2016; Interview GIZ, November, 2016.
[25] Interview MST, November, 2016; Interview HST, December, 2016.
[26] Interview HCC, December, 2016.

Table 3. Summary of partners in municipalities[a].

	Partner from municipal development plan	Partners identified to finance infrastructure in municipal development plan	Comments
Houndé	GIZ Albacete, Spain Saint Fons, France	Burkina Manganese US$ 473,824 (new water tower) ADAE US$ 3,313.45 (expansion of water tower system)	GIZ stated to only provide capacity building for the municipality though they did supply financing for a water tower and standpipes in 2012
Komsilga	Friendship Japan Charles DuFour Kinder-In-Not Canadian International Development Agency	Decentralized Partners (does not specify which partner) US$ 223,658	The municipal accountant stated that there were two NGOs also partnered with the commune, ACT and ASMT
Banfora	Oxfam Quebec Chauvigny, France Trino, Italy Sikasso, Mali Geisenhein, Germany Kona, Japan Ouahigouya, Burkina Faso Pouytenga, Burkina Faso	Oxfam Quebec: expansion of water tower system to two villages (Siniena and Tengrela); eight standpipes	Oxfam Quebec was the most noted partner of the municipality for infrastructure development
Moussodougou	Oxfam Quebec ACAD – City of Mondonville	ACAD – new water tower in Diamon	ACAD was stated to be the primary partner of the municipality, which had realized a water tower in a village. The service technician highlighted that receiving transfers from the government was unlikely and that partner finance was preferable

Sources: Commune de Banfora & Oxfam Quebec, 2013; Commune de Hounde & GIZ, 2009; Commune de Komsilga & GRAD Consulting Group, 2013; Commune de Moussodougou *et al.*, 2016a; primary data.
[a]Interviews HCC, December, 2016; HSG, December, 2016; HST, December, 2016; KCC, December, 2016; KSG, December, 2016; KST, December, 2016; BCC , November, 2016; BSG, November, 2016; BST1, November, 2016; MSG, November, 2016; MST, November, 2016).

partner finance relationships, stating that it would be necessary to find another such partnership and that they would be lost without their partner[27].

4.2. *Project finance*

'They just pass by the mayor's office.'
Secretary General in Komsilga

Project finance relates to infrastructural development projects financed by donor agencies. Rather than contributing to development of the municipality, this type of aid focuses only on individual infrastructure projects. Project finance most often does not adhere to the municipal development plans for water

[27] Interview MST, November, 2016; Interview HST, December, 2016.

infrastructure development. In other words, the municipal plan for development of water infrastructure is not a basis on which these development projects are selected and implemented. This finding is supported by a WaterAid study of public financing in the Sub-Saharan water services sector. It finds that '[s]ignificant amounts of donor-funding and Non-Governmental Organization (NGO) funding […] are off-budget, targeted directly at areas without being reported in Ministry of Finance budgets and spending' (WaterAid, 2014:9). As project finance largely occurs with minimal involvement of the municipality estimates of the resources mobilized for project finance in the towns were not available from the municipality. In the four municipalities studied, the only involvement of the municipality was that they were informed of the location of the infrastructure developed under the project[28]. The service technician in Houndé expressed concern that he did not know what or where infrastructure was being constructed in the municipality and that those involved with these projects viewed informing the municipality as a mere formality[29].

The development of water infrastructure, which is not in line with the municipal development plan highlights that donor priorities are not necessarily aligned with priorities for infrastructure development identified by the municipality. Moreover, project finance does not appear to be concentrated in the municipalities with the greatest need. The municipalities with the highest level of access also had the greatest evidence of project finance. Specifically, Komsilga and Banfora (with access rates of 99.4 percent and 75.4 percent, respectively) also have the highest levels of project finance (see Table 4). Illustrative of the misalignment of project finance with the needs of municipalities is the development of wells in Komsilga. Whereas the municipality had allocated resources for the maintenance of seven wells and the municipal development plan targeted the development of 30 wells and kiosks, a total of 74 wells was developed in the municipality over the past two years. This means that twice the amount of planned water infrastructure was constructed in Komsilga.

5. Discussion: some towns are more equal than others

'To succeed, you need to demonstrate success and dissociate yourself from failure.'
Rap (2006)

Given the urban bias in public funding of water infrastructure and the limited local government funding of water infrastructure, each of the four municipalities showed a strong dependency on donor funding for water infrastructure development. This donor finance was accessed either through partnerships with donor agencies and foreign municipalities or through donor-driven project funding. The strong dependency on donor finance, however, is problematic for a number of reasons.

First of all, the reliance on project finance leads to the development of water infrastructure outside the scope of public accountability. Whereas municipal development plans may be subject to public scrutiny, project finance which contributes to the development of infrastructure outside the municipal development plan is decided upon outside the public domains. Between one-third and half of the villages in three municipalities showed evidence of project finance in which the municipality was only minimally

[28] Interview HST, December, 2016; Interview KSG, December, 2016.
[29] Interview HST, December, 2016.

Table 4. Evidence of project finance.

	Houndé	Komsilga	Banfora
Villages with infrastructure development that is not stipulated in the municipal development plan	5 out of 15	17 out of 36	8 out of 22
Percent of villages with project finance	33.33%	47.22%	36.36%
Level of access	63.8%	99.4%	75.4%

Sources: Commune de Banfora & Oxfam Quebec, 2013; Commune de Hounde & GIZ, 2009; Commune de Komsilga & GRAD Consulting Group, 2013; INSD & MEA, 2015.

involved. Earlier research suggests that NGOs often have incentives to follow the agenda of the government donor and to focus on short-term and quantitative outputs, rather than on the communities they aim to serve (Edwards & Hulme, 1996; Hickey, 2002; Mitlin *et al.*, 2007; Rusca & Schwartz, 2012). Rusca & Schwartz (2012) suggest that international NGOs 'may have difficulties reconciling 'upward' accountability to the funding agency and 'downward' accountability to the beneficiaries and southern partners with which they work'. This leads to the concern how evenly project finance is divided over different municipalities.

Apart from possible inequalities between villages within the same municipality, donor-funding also leads to questions between villages in different municipalities. As highlighted, not all municipalities are equally able to attract project finance. Komsilga was the site for many donor-funded interventions. So what makes Komsilga more attractive for donors than the other municipalities? An important explanation for the popularity of Komsilga appears to be that it has the characteristics that allow donors to relatively easily develop and implement project interventions. Donors have an interest in ensuring that their water infrastructure interventions are successful and the selection of suitable project site can greatly aid the chances of a successful intervention (Rap, 2006). This is echoed by a WaterAid publication that argues that donors 'delegate decision-making to domestic partners, who in turn choose easier options to increase the likelihood of success, as well as keep costs lower than the hardest to reach and most marginalised communities' (WaterAid, 2014:12). In our case studies, Komsilga exhibits a combination of factors, which make it a preferable site for such interventions. These factors include its rural characteristics, the developed project-management capacity, an established record of previous successes, and a well developed network of informal relationships. These factors are elaborated upon below.

Komsilga is a preferred site for interventions because it has rural characteristics whilst also being in close proximity to the capital. At a national level, the success of ONEA and the achievement of the MDGs in urban areas mean that donors are targeting rural areas for interventions. Accordingly, Komsilga's rural characteristics and proximity to the capital mean that it is a good site for testing interventions for rural contexts and facilitates multiple visits by evaluators, which donors require to evaluate their interventions.

As donors choose Komsilga to test their interventions, both the municipality and community-based organizations within the municipality develop capacity to manage and adhere to project management and financial management requirements associated with donor funding. Komsilga's Municipal Development Plan, for example, demonstrates an aptitude for managing these other sources of finance. Komsilga's plan also assigned partners at the village level, which demonstrates that the villages have developed sufficient capacity to manage partner finances without support from the municipality. This level of sophistication indicates that Komsilga and its villages have developed their capacity to working with external sources of financing. This is important as Edwards *et al.* (1999:130) argue, 'even where they work through 'partners', many [donors] tend to dictate the scope and pace of work through their control over funding and procedures'. This means that local partners need to have the capacity to account for funding and adhere to the procedures stipulated by the donors who emphasize the 'capacity for accountability' as constituting an important element for ensuring a 'successful' project (Rusca & Schwartz, 2012). What results is a reinforcing apparatus in which donor-funded projects lead the municipality and community-based organizations in Komsilga to increase their capacity to accommodate donor-funded projects. This, in turn, leads to more donor-funded projects. Organizations in the other municipalities, which have less capacity are unable to attract donor-funded projects that will allow them to develop such capacity. As a result, they are stuck in a low-level equilibrium which they have difficulty getting out of.

Apart from this more formal project management capacity, Komsilga also benefits from the existence of informal networks with the central government. The previous mayor of Komsilga worked at the General Directorate for Potable Water, during his tenure as mayor[30]. Small municipalities often struggle to gain the attention at the national level, as they lack the power and influence of large urban centers. Therefore, informal links such as between the mayor and the General Directorate may act to increase Komsilga's relative influence on the national government. For example, the government selected Komsilga to be a test site for a private operator contract with SAWES[31]. This informal network is helpful for Komsilga as project proponents might seek the advice of the Ministry of Water and Sanitation to identify pilot sites for their interventions. Having a mayor working at the DGEP facilitates opportunities to direct projects to that mayor's municipality[32].

Although the access to project finance of Komsilga may be beneficial for the municipality, it also comes at a cost. The national government does not transfer public resources transferred due to the low capacity at the municipal level to manage these funds. As a result, the municipality's reliance on these other sources of financing develops a parallel structure to public finance, which is used to finance their water infrastructure. Komsilga, thus, develops capacity to manage the requirements of partner and project finance, instead of developing public financial management capacity. The development of capacity in the parallel structure has long-term implications for the municipality and its use of public finance. Over time the municipality's ability to manage donor funds increases whilst its access (and ability to manage) public funds remains limited. Moreover, it does little to prioritize municipal budget allocations for water infrastructure development. This reinforces the already strong dependency on donor funding. This vicious circle was both referred to by a representative of the General Directorate of Water as well as a representative of Komsilga. However, both approached the circle with alternative solutions. Whereas the representative of the General Directorate emphasized the need for the municipality to develop capacity in order to obtain national government funding, the representative of the municipality emphasized the need to transfer national government funds in order to develop capacity.

[30] Interview KAM, December, 2016.
[31] Interview KSG, December, 2016.
[32] Interview KAM, December, 2016.

6. Conclusion

This analysis of the public finance system in Burkina Faso reveals interconnected biases. Funding mobilized for the 2007–2015 National Program is only minimally allocated for rural or small town water infrastructure development, whereas the transfers to ONEA were almost twice the combined amount of transfers to regions and municipalities, and nearly quadruple the amount when adjusted by population served. This indicates an urban bias, which is alarming for two reasons. First, the urban sector already has relatively-high levels of access to water, and therefore allocations to the urban sector do little to work towards distributive equity. Second, the contract between the national government and ONEA allows ONEA to reinvest resources into their service areas. This contractual stipulation means that these transfers reinforce the existing inequities by preserving the continual investment of financial resources only in the urban areas. During implementation of the 2007–2015 National Program, the Centre region received the most resources to develop their water infrastructure. This is problematic because the Centre region already had the highest level of access to water in Burkina Faso. The other regions analyzed for this thesis, Hauts-Bassins and Cascades, are among the regions with the lowest levels of access to water, and yet received 16.9 percent and 18.5 percent fewer resources than the Centre region, respectively[33]. This demonstrates that the public finance system in Burkina Faso is currently unable to either achieve distributive equity or to target investment to the areas with the lowest level of access.

At the same time, the local government also does not appear to view it as a responsibility to fund water infrastructure development from its municipal budget. Only one of the four municipalities allocated a very small amount from its municipal budget for infrastructure maintenance. Rather, the municipalities appear to view infrastructure financing as a responsibility of partners and donors. The dependency on donor-funding, however, is also problematic. Municipalities that have favourable characteristics for successful project implementation appear to be at an advantage of sourcing these funds. As a result, municipalities that are already relatively well served, receive a disproportionate share of these donor funds as this funding is subject to the priorities and interests of the donors, which may not be aligned with distributive equity. In the case study municipalities, these other sources of funding were not geared towards the municipalities with the greatest need. Komsilga had the highest level of access, but as discussed above possesses a combination of attractive attributes that allow it to easily access project funding. Some small towns are thus confronted with a double bias. First an urban bias in which the majority of public finance goes to urban centres served by ONEA. Secondly, by a donor-bias in which some towns are favoured for project implementation due to favourable site characteristics.

References

Commune de Banfora (2016). *Compte Administratif Gestion Budgetaire 2015.* Commune de Banfora, Banfora.

Commune de Banfora, & Oxfam Quebec (2013). *Plan Communal de Development Sectoriel d'Approvisionnement en Eau Potable et d'Assainissement (PCD-AEPA) 2013-2018.* Commune de Banfora, Banfora.

Commune de Hounde (2016). *Budget Primitif Gestion 2016.* Commune de Hounde, Hounde.

[33] When the regional data are corrected for rural population, the disparities between regions grow, with the Centre Region receiving over three times the resources as the Cascades and Hauts-Bassins regions combined (INSD & MEA, 2015; MEA, 2016; OANDA, 2016; World Bank, 2016).

Commune de Hounde & GIZ (2009). *Plan Communal de Development Sectoriel d'Approvisionnement en Eau Potable et Assainissement 2010–2015*. Commune de Hounde, Hounde.

Commune de Komsilga (2016). *Budget Primitif Gestion*. Commune de Komsilga, Komsilga.

Commune de Komsilga & GRAD Consulting Group (2013). *Plan Communal de Developpement Commune Rurale de Komsilga*. Commune de Komsilga, Komsilga.

Commune de Moussodougou, IGIP Afrique & Oxfam Quebec (2013). *Rapport Diagnostic Approvisionnement en Eau Potable et Assainissement: Version Provisoire*. Commune de Moussodougou, Moussodougou.

Commune de Moussodougou, Oxfam Quebec & IRC International Water and Sanitation Centre (2016a). Plan d'Action AEPHA 2016: Commune de Moussodougou.

Commune de Moussodougou (2016b). Budget Primitif Gestion. Moussodougou.

DGEP (2015). *Programme National d'Approvisionnement en Eau Potable à l'Horizon 2030*. DGEP, Ouagadougou.

DGRE (2006). *Programme National d'Approvisionnement en Eau Potable et d'Assainissement à l'horizon 2015*. DGRE, Ouagadougou.

Edwards, M. & Hulme, D. (1996). Too close for comfort: NGOs, the state and donors. *World Development 24*(6), 961–973.

Edwards, M., Hulme, D. & Wallace, T. (1999). NGOs in a global future: marrying local delivery to worldwide leverage. *Public Administration and Development 19*, 117–136.

Hickey, S. (2002). Transnational NGOs and participatory forms of right-based development: converging with the local politics of citizenship in Cameroon. *Journal of International Development 14*, 841–857.

Humphreys, E., van der Kerk, A. & Fonseca, C. (2018). Public finance for water infrastructure development and its practical challenges for small towns. *Water Policy 20*(s1), 100–111.

INSD (2016). *Annuaire Statistique 2015, Ministère de l'économie des finances et du développement*. Institut National de la statistique et de la démographie, Ouagadougou.

INSD & MEA (2015). *Resultats INO 2015 Fevrier, Ministère de l'économie des finances et du développement*. Institut National de la statistique et de la démographie, Ouagadougou.

MAHRH (2006). *Programme d'Application de la Réforme: Le système de gestion promu par la Réforme*. Ministère de l'Agriculture, de l'Hydraulique et des Ressources Halieutiques, Ouagadougou.

MEA (2016). *Revue des Dépenses Publiques de 2007 à 2015 dans le Secteur Eau et Assainissement: Annexes*. Ministère de l'Eau et de l'Assainissement, Ouagadougou.

MEA & DGESS (2016). *Rapport de Performance de l'Annee 2015 du Secteur Eau et Assainissement*. Ministère de l'Eau et de l'Assainissement, Ouagadougou.

Mitlin, D., Hickey, S. & Bebbington, A. (2007). Reclaiming development? NGOs and the challenge of alternatives. *World Development 35*(10), 1699–1720.

OANDA (2016). Currency Converter, Average. Retrieved from https://www.oanda.com/currency/average. Accessed December 8, 2016.

ONEA (2016). *Carte d'Implantation*, Retrieved from http://oneabf.com/carte-dimplantation/. Accessed March 2, 2017.

Pietila, P.E. (2006). Role of municipalities in water services. Tampere University of Technology Publication, Tampere, FL, 617.

Rap, E. (2006). The success of a policy model: irrigation management transfer in Mexico. *Journal of Development Studies 42*(8), 1301–1324.

Rusca, M. & Schwartz, K. (2012). Divergent sources of legitimacy: a case study of international NGOs in the water services sector in Lilongwe and Maputo. *Journal of Southern African Studies 38*(3), 681–697.

van den Bergh, C. & Danilenko, A. (2017). *Performance of Water Utilities in Africa*. World Bank, Washington, DC.

WaterAid (2014). *From Promise to Reality: The Urgent Need for Southern African Leaders to Deliver on Their Water, Sanitation and Hygiene Commitments*. Retrieved from http://www.wateraid.org/what-we-do/our-approach/research-and-publications/view-publication?id=af60d21e-3493-4b6e-b30c-98633054ac24. Accessed August 22, 2017.

WHO/UNICEF JMP (2015). *Progress on Sanitation and Drinking Water–2015 Update and MDG Assessment*. World Health Organization, Geneva.

WHO/UNICEF JMP (n.d.). *Improved and Unimproved Water Sources and Sanitation Facilities*. Retrieved from https://www.wssinfo.org/definitions-methods/watsan-categories/. Accessed August 10, 2017.

World Bank (2016). *Burkina Faso Data*. Retrieved from: http://data.worldbank.org/country/burkina-faso. Accessed July 1, 2017.

WSP (2011). *Water Supply and Sanitation in Burkina Faso: Turning Finance Into Services for 2015 and Beyond*. AMCOW Country Status Overview. WSP, Nairobi.

Portrait of a successful small-town water service provider in Nepal's changing landscape

Sanna-Leena Rautanen[a] and Pamela White[b]

[a]*Corresponding author. RWSSP-WN, FCG International, PO Box 950, Helsinki, Finland. E-mail: sannaleenar@gmail.com*
[b]*FCG International and University of Helsinki, PO Box 3, Finland*

Abstract

This study was made in Nepal's Tarai plains, where rapid population growth over the past decade has transformed a large number of rural bazaars and roadside hubs into vibrant small towns. This study draws a portrait of a distinctly successful small-town water supply scheme and its service provider, the Murgia Water Users and Sanitation Association. Exploring this particular case with regards to social, technological, financial and organisational systems, and by comparing the performance of this case against 63 other water service providers in Nepal, the study asks: how could there be more of this type of successful water service provider? This scheme was constructed during the bilateral Rural Water Supply and Sanitation Support Programme Phase III, Nepal-Finland cooperation (1999–2005), using the typical rural approach, namely community management, with strong capacity building. Since then the service modality in this study case has evolved towards a professional community-managed service delivery. The success is rooted in good water governance principles: participation, responsiveness, financial transparency, accountability and overall strong commitment and vision, as well as strong technical assistance. They have resulted in re-investment in both the capital maintenance expenditure and into new infrastructure, even into an entirely new water supply scheme.

Keywords: Capacity; Nepal; Small towns; Technical assistance; Water services

Introduction

This study takes the approach that water and sanitation service provision encompasses social, technological, financial and organisational systems which are strongly interdependent and which together form a specific service delivery system. It considers an 'in-between' settlement in Nepal's Tarai plains, where the rapid population growth over the past decade has transformed a large number of small rural bazaars and roadside hubs into vibrant small towns. The water supply schemes that were constructed as rural systems some 15–20 years have a rather different operational environment now. This case study represents one successful water service provider that has managed to evolve from

doi: 10.2166/wp.2018.006

rural to professionally operating small towns' service provider. The question is – how to have more water service providers such as the one being studied? Can this case be replicated?

Nepal made good progress in achieving its Millennium Development Goals, and met its target for water supply. In urban areas the water supply coverage is now 90.9% and in the rural areas 91.8% (WHO/UNICEF, 2015). However, even if the water supply coverage in urban areas appears high, the service levels are generally poor. A nationwide functionality study concluded that out of 41,205 piped water supply schemes covered, 68.2% provided whole-year supply and only 25.4% were described as '*well-functioning*' (Government of Nepal, 2014a). The Urban Water Supply and Sanitation Sector Policy 2014 outlined that '*In the cities and towns water supply coverage is poor; supply pressure is low; supply is intermittent; water is not potable without further measures at household level to make it potable; non-revenue water is high; consumer satisfaction is low. All urban water supply services are in need of substantive improvements*' (Government of Nepal, 2014b). The WASH sector status report attributes the shortcomings in the sector to several factors ranging from institutional weakness and reliance on donor money, to unsustainable practices that disregard rehabilitation and maintenance works, or pay little attention to environmental impacts (Government of Nepal, 2011).

Nepal has recently launched its Water Sector Development Plan (SDP) for the period of 2016–2030. This plan identifies the National Water Supply Corporation as the operator for the urban water supply while Water Users and Sanitation Committee (WUSC) are the operators of small towns and village systems. While there is a 1,000 population cut-off point for whether the water supply schemes should fall under the Ministry of Water and Sanitation or the Ministry of Federal Affairs and Local Development, the same cut-off does not seem to apply to the water service provider. Therefore, WUSCs can be found from peri-urban and even in urban areas. SDP gives WUSCs several overlapping roles as '*the operator, service provider, regulator, monitor, fund raiser and manager*', calling for WUSCs to include at least 33% representation of women and representation from disadvantaged groups (Government of Nepal, 2016). Top-down planning and implementation is still evident in many urban and semi-urban locations where there is a large population to be served. In these cases, the planning and implementation principles applied tend to focus on engineering standards, while the ultimate operator, the water service provider itself, gets less attention. This can lead to dysfunctional operators.

The purpose of this study is to draw a portrait of a successful small town water supply service provider, compare it to 63 other water service providers in Nepal, and explore to what extent and why it has been successful. In the last two decades, community-based management, through Community-Based Organisations (CBOs) became the widespread and most common model for management of rural and village water supply (Moriarty *et al.*, 2002). CBOs for water services have been promoted in many countries as they provide a structure that enhances ownership and empowers the recipient community. However, the successful implementation of these models also proved challenging and very much context dependent. It has been documented that for this model to work, a strong support from governmental forms or NGOs is required (Moriarty *et al.*, 2002). However, the imposition of improved commercial practices also raises criticism as it is potentially abusing the initial design of the CBOs establishment which were originally established as organisations 'of the community, by the community, and for the community' (WSP, 2011, p. 6).

In the following sections we present a comprehensive description of the study area of Parroha and the local water provider, the Murgia Water Users and Sanitation Association (WUSA) and its Water Users and Sanitation Committee (WUSC) as WUSA's executive body. In our analysis we compare the performance of our case study against other operations elsewhere in the country and provide a detailed description of how this performance can be explained in terms of social, technological and financial and organisational arrangements. We propose certain adaptations that could potentially pave the way to the implementation of sustainable services.

The study area

Until recent local elections, the Village Development Committee (VDC) was the lowest tier of local governance in Nepal. Parroha VDC was a roadside market area located some 15 km from Butwal municipality, in the Western Development Region of Nepal. Butwal is a major market and transit hub since ancient times, given its strategic location connecting trade routes from north to south, and from east to west. Today the east–west highway running through the entire country is the biggest transportation corridor in Nepal. Parroha represents a typical fast growing bazaar area along this highway. The Tarai has the most important agricultural lands in Nepal, and the interest in groundwater irrigation has increased significantly over past decades with a number of donor-funded irrigation projects exploring both shallow and deep tube wells for irrigation (Sugden, 2014). The number of households has doubled in 15 years and the number of businesses has been growing and diversifying at equal pace. In 2001 Parroha VDC had already a total population of 19,055 (Government of Nepal, 2001) which made it a small or intermediate town by any definition already at that time (Figure 1). Parroha represents a typical 'in-between' settlement that has both rural and urban characteristics. The majority of households still rely on crop cultivation and livestock, even if an increasing number of people are involved in various businesses in the secondary or tertiary sectors (not agriculture), whether in Sainamaina municipality itself or in nearby Butwal.

Parroha VDC merged with the neighbouring Dudharakchhe VDC in May 2014, to form Sainamaina municipality[1]. At the time of the 2011 Nepal census, Sainamaina Municipality had a population of 48,178 (Government of Nepal, 2011). Nevertheless, the approach taken for construction of the water supply scheme in Parroha 15 years ago was that of a community-managed rural water supply.

Murgia WUSA is the water service provider in this study. It is the institution that delivers water to the users and is responsible for the day-to-day operations, from maintenance to administration. The operational body organised in the Murgia WUSC is an elected group of community members that manages the water network system, via community management (Lockwood & Smits, 2011).

Parroha is located in Nepal's southern belt, a plain area referred to as 'Tarai'. The area is generally regarded as having vast groundwater reserves, and millions of households depend on shallow tube well hand pumps for their domestic water. While the basin where Parroha is located is part of the broader groundwater regime of the overall Ganges Basin there is considerable local variation over relatively short distances. In Parroha these local hydrological conditions do not favour shallow tube wells, but demand deep tube wells (typical of northern Tarai). Ecological sustainability is hence a locality-specific question that is largely unresolved in Nepal, while the demand for irrigation water grows as quickly as domestic water aspirations in new urban areas.

Murgia WUSC operates a piped water network, using groundwater lifted into an overhead tank. This technology is not a typical *rural* water supply system in Nepal, but rather something that has become very typical for the urban and small-towns areas of the country. The 63 water service providers from

---

[1] For the purposes of this article we will continue to refer to the area as Parroha.

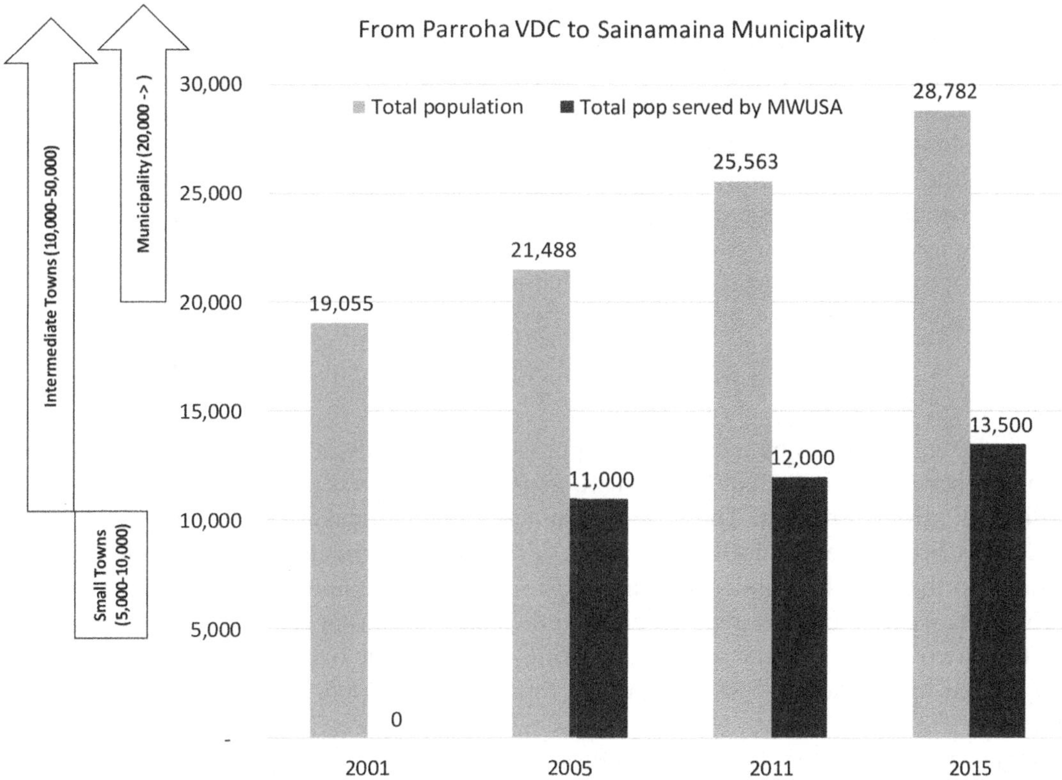

Fig. 1. Parroha VDC, Rupandehi district – changing in time. *Sources:* Government of Nepal, 2001; Government of Nepal, 2009; Personal Communication Murgia WUSA.

Nepal that we compare to include both water supply users' committees and associations, as well as water supply management boards and branches of the Nepal Water Supply Corporation.

The water supply scheme operated by this service provider was constructed under the Rural Water Supply and Sanitation Support Programme Phase III (RWSSSP III), with support from Nepal-Finland bilateral cooperation, which was active in the Lumbini zone of Nepal from 1999–2005. Under this programme, schemes were developed using a participatory interactive community management approach that was backed up with comprehensive step-by-step capacity development and close monitoring. The Finnish bilateral development support differed from many other sector donors, with international and Nepali technical advisers providing hands-on support to local government and WUSCs. This has ensured WUSCs' capacity development and institutional strengthening through learning-by-doing and targeted training at grassroots level. White *et al.,* (2015) described how this same Finnish method resulted in much greater functionality of rural water supply in far west Nepal, when compared with schemes constructed within the usual centralised government system (White *et al.*, 2015).

The Completion Report of RWSSSP noted the importance of capacity development in the rapidly growing communities: '*Particularly relevant in the large schemes by the road corridor involving overhead tanks, large gravity flow systems and one water treatment plant mean increased consumer expectations for higher levels of service and water quality (...)*' (RWSSSP, 2005). Rautanen (2016) developed the ideas of several authors to consider the critical pre-project and post-project issues influencing the sustainability of rural water schemes. This paper demonstrates how these factors were addressed in Murgia, even as it crossed over from rural to small town setting.

In 1993, the area consisted of clusters of households sharing water sources locally, in this case usually hand pumps or open water sources on a self-help basis. In the year 2014, map the urban characteristics are manifested along the east–west highway corridor with dense housing. In the early 1990s there were only two roads, the east–west highway and another gravel road. The rest of the road network consisted of small earthen paths that connected the housing clusters to each other. In 2016 the road infrastructure had significantly improved. Electricity and water were available and the municipality declared as 'Open Defecation Free', moving fast towards Total Sanitation.

Methodology

The methodology is guided by a case study approach. Case study is an empirical inquiry that investigates a phenomenon within its real-life context, usually being based on an in-depth investigation of a single individual, group, or event. The information about this particular case – the Murgia WUSA and its WUSC – has been collected through interviews with key informants, site observation and an interaction workshop that could be understood as a focus group discussion. Murgia WUSC speaks in this paper through its own voice. The report builds further on monitoring reports and field notes made in 2013–2016 by two of the authors while monitoring other similar overhead tank water systems. The data were backed up by the RWSSSP III project reports (2000–2005). The financial figures originate from Murgia WUSC itself.

Case studies have sometimes been criticised for having a bias toward verification, understood as a tendency to confirm the researcher's preconceived notions (Flyvberg, 2006). In this study, this is a valid concern given one author's long-term engagement with the case. Yet, the objective is not to portray solely this case as a success story – which is already the case – but rather to find plausible interventions and practices that could explain the success and provide greater detail on these practices. The differences between the case study and other operators are derived from a comparison between the performances of 63 other Water Service Providers in Nepal against the Nepal Service Level indicators (Table 2). Furthermore, the data collection for this study included a random selection of consumers within the service area to verify that the perception from the reported performance figures match with the perceptions on the ground. There is an increasing demand for private connections and service level improvements, as well as lift schemes that call more professional management than what is typically practised in the rural water systems, hence any learning from a successful case will be useful for other WUSCs. Below we present the comparison data.

What is the performance of the Murgia WUSA compared with 63 other small towns' water service providers in Nepal?

Using the Service Level categorisation as practised in Nepal (Government of Nepal, 2016), it is evident that the Murgia WUSA water services score 'High' against most of the service level indicators. With supply of 148 litres per capita per day, the Murgia WUSA is almost at the 'high' category (starting at 150 l/capita/day). The water is sufficient for hygiene needs. The Murgia WUSC tests water quality at a certified laboratory twice per year: once during the rainy season, and once during the winter season. All consumers have a private connection, and four public tap stands serve public needs. The reliability is high, with all day service. Any interruptions are mainly caused by maintenance needs, including periodic emptying of the overhead tank for cleaning purposes. Services are usually restored within a day. All these aspects rank 'high' against the water service level indicators in Nepal.

The new service level indicators include a quarterly consumer satisfaction survey. This has not been done in Murgia, but given the low official complaint rate (none over the past year) and the fact that the consumers are paying their water tariffs, we deduce that the consumers are satisfied with the services. There is also a suggestions box available by the water tariff payment counter, another way of filing a complaint. Since the office manager is available in the compound daily, the users are welcome to contact him directly with any issues. An official complaint costs NPR 10 (USD 14 cents) which is unlikely to prevent customers filing a complaint should they have a reason to do so. If the Murgia WUSA goes ahead with its SMS plans, the same tool could be used for quarterly consumer surveys as well.

Table 1 shows how Murgia WUSA compares to 63 other small towns' water supply providers (WSPs). The ranking follows the order of the *Water Service Providers Data Book, 2070–2071 (2013–2014)* (Government of Nepal, 2015). These include both the government mandated utilities as well as WUSAs operating on the same principles as Murgia. It is clear that Murgia scores better than average on most indicators. The performance is remarkable both in terms of coverage rates, as well as in economic and financial performance.

Table 1. Murgia WUSA compared against 63 other water service providers in Nepal.

Indicators	Unit	Murgia WUSA*	Average of 63 WSPs**	Murgia WUSA*, ** position out of 63 WSPs
		(07/2016)	(07/2014)	*, **
Water supply coverage = [population served with water supply] × 100 / [total population in the service area]	%	90%	63.4%	9
Consumption per capita = [total annual water sold (m^3) × 1,000/365] / [number of people served]	Litres/capita/ day	148.1	70.4	3
Connections metered	%	100%	93.40%	n.a.
New connection fee (residential, new connections)	NPR	14,500	10,661	18
Monthly bill per connection	NPR/ household/m^3	221.2	n.a.	30
Average tariff per m^3 = [total annual billing (NPR)] / [total annual consumption (m^3)]	NPR/ m^3	8.4	15.4	56
Unit production cost = [annual O&M cost (NPR)] / [total annual production (m^3)]	NPR/m^3	4.3	n.a.	3
Operating ratio = [Total annual operation & maintenance cost (NPR) / total annual billing (NPR)]	%	54.6%	102%	8
Staff per 1,000 connections = [number of utility staff] / [number of utility connections/1,000]	Ratio	3	7.2	4

Sources: *Murgia WUSA communication September 9, 2016; **Government of Nepal (2015).

Results in social, organisational and financial terms – are they replicable?

The original plan was to construct a 50,000 litre capacity overhead tank to serve 756 households, pumping from one deep well. In 2004 the operation started, serving 745 households with a population of 4,500. The number of connections increased rapidly within the first year of operation. After more than 10 years of operation, the system serves 2,317 households (i.e. more than three times the original plan). Table 2 shows how various operational aspects have changed from the first year of operation compared to the fiscal year 2015/2016. The new boreholes that the Murgia WUSA financed from their own sources have made the increased production possible (even if one of the boreholes has dried up). From the technological point of view there is nothing different to what other water service providers have – this aspect is fully replicable.

Murgia WUSA has managed to establish a performance that over 10 years after commencement of operations has followed through with the indicative positive evolution registered in the initial years (2004–2006). In Murgia they have augmented the quantity of water supplied, not only in terms of population that increased by 300%, but also in water available and consumed per capita (litres/capita/day) that increased by 48%. They have also been able to become more efficient as it is distilled from the staff per 1,000 connections ratio. We provide an analysis of the organisational set up, the financial arrangements and the social aspects related to functioning of the model that help explain the sustainability of these figures.

Organisational set-up – is this replicable?

Murgia WUSA was established in 2001 in Parroha VDC. While WUSA is the executive body, the WUSC is its operational committee. Main WUSC has 11 members (four female and seven male).

Table 2. Murgia WUSA and its services 2004–2016.

Indicator	Unit	2004/2005 1st year of operation	2005/2006 2nd year of operation	2015/2016 Most recent year closing
Staff	number	3	5	7
Staff per 1,000 connections	number	4.0	5.8	3.0
Population served with water supply	population	4,500	5,500	13,500
	households	745	856	2,317
Total population in the service area	population	10,000	11,000	15,000
	households	1,656	1,712	2,574
Water supply coverage	%	45%	50%	90%
Number of private connections*	number	745	856	2,317
Connections metered	%	100%	100%	100%
Total billed consumption	m^3	164,250	219,000	730,000
Consumption per capita served	litres/capita/day	100	109	148
Production per capita	m^3/day/capita	0.10	0.11	0.15

Source: Murgia WUSA, personal communications September 9, 2016.

*Other connections: four public tap stands and nine institutional taps (5 schools, 1 police office, 1 health post, 1 temple, 1 local government office.

Although it has a few paid staff members, the main functioning is supported by volunteer staff. The members are elected during the General Assembly every three years. These are not paid positions. Murgia WUSC has the following three sub-committees for operational tasks, with one WUSC member as the coordinator in each, with members nominated by the WUSC itself. (1) Procurement Committee which conducts market surveys, collects quotations, and provides instructions for the employees on what to purchase. In this way they guarantee that even the very minor day-to-day purchases remain transparent. (2) Repair and Maintenance Committee that plans the minor repair and maintenance. (3) Construction Committee that is set up when a new major construction is started, such as the new toilet-bathroom block and related compound improvements over the past year.

Apart from that, Murgia WUSA has the following other committees also based on volunteering manpower but external to the WUSC. (1) Advisory Committee that is nominated (as opposed to the WUSC which is elected) during the General Assembly when the new WUSC is elected (every three years). The members are invited to the WUSC meeting when there are relevant items in the agenda, such as the meeting before the General Assembly when preparing the next year's budget and programme, or when preparing proposals for new programmes. (2) Internal Audit Committee whose members are nominated during the General Assembly and cannot include WUSC members. Its task is to audit the income and expenditure every three months. At the beginning, there were also 16 small cluster-wise subcommittees, consisting of three members only. These operated independently with their own chairperson, who was the first contact point for such issues like new connections within the cluster. At the beginning these subcommittees were needed to localise the community mobilisation and consumer relations when applications for the connections started to arrive from various parts of the network. At present, these subcommittees are not active anymore. Instead, a WUSC member has been assigned as the first contact point in their locality.

The WUSC meets on the '*8th of every month*'. It is a non-negotiable event that has to take place at the given time. In this meeting, the Office Manager presents the monthly income and expenditure, among other deemed relevant things, and gives an opportunity for the WUSC members to respond to any critical issues in a timely fashion. The regular meeting practice makes the operation very transparent as the figures are shared systematically for the entire WUSC on a regular basis. These are recorded in the WUSC minutes, which in turn are available to anyone interested. The flow of information among the community has proven to be one of the explanatory elements to their success. However, the permission to view the minutes is granted by WUSC, not by the staff members such as the office manager. The WUSC treasurer has an active role as she closes the accounts every evening by signing the summary of the day's income and expenditure. The monthly statement is based on these signed results, and is open to all interested people.

Initially Murgia WUSA had three full-time, paid staff, growing into five persons during the second year of operation when the number of connections started rising. Fifteen years later the staffing consists of seven full time employees: office manager; sub-accountant; plumber; meter reader; operator; guard and office assistant. The Murgia WUSC members are proud of their staff and their team spirit is strong. The present WUSC chairperson described the seven full-time staff as '*loyal, efficient, and skilled, with strong team spirit, responding efficiently to customers' needs*', and said each team member is ready to perform any task: '*even the office manager will do meter reading if needed*' (personal communication). Other WUSC members are also active, the garden around the overhead tank and office compound being one visible flourishing manifestation of their commitment and interest in maintaining the immediate environment.

The office manager considers that providing water is 'holy work', given that water itself is holy. They are proud to be members and staff of the Murgia WUSA. Being associated with this institution is esteemed. A WUSC member reported in September 2016 that *'there are several chairpersons and presidents in a number of committees and institutions in this locality. Yet, the position in this WUSC is seen as the most prestigious'* (personal communication). The authors assume that the reputation builds on the excellent track record in terms of reliable water services, transparency and accountability. The position is highly appreciated even if the only 'income' to be earned is the meeting allowance.

The functioning of the organisational structure of Murgia WUSA relies heavily on local social arrangements, and local pride and ownership. In this way, we consider that the organisational is theoretically replicable, but that its actual implementation will be better understood as part and parcel of the social considerations attached to the model.

We consider this organisational set-up replicable.

Social considerations – is this replicable?

The Murgia WUSC appears to be committed and motivated to provide sustainable high quality water services, while continuously attempting to improve and extend the operations. As we have seen earlier on, the commitment translates from a sense of pride in delivering a good service. There is a strong focus on representing and serving all community members, using principles of gender equality and social inclusion that were inculcated during the establishment of the scheme. This was an element emphasised by the project. One of the internal drivers that keeps Murgia WUSC moving forward, seeking new and improved ways of operations and services, appears to be the notion of 'modernity' – the emerging small town represents something new, something modern, something wealthy and promising, an opportunity. Water (including piped water to the household) and sanitation are important aspects of this type of 'modern well-being.'

In the Tarai, migrants have arrived from both hill communities and from India, and there is a more diverse mix of religions, ethnicities, languages and castes than in rural hill villages. The social order of hill villages is diluted and the priorities for water use of farmers and non-farming populations vary. The slower progress in achieving national sanitation goals in the Tarai seems to be evidence, or a result, of less cohesion. Exclusion in the past led to the civil war, and very recently to strikes and violence. Social inclusion in Murgia has proven key to ensure a functional system. If groups are excluded, as is the case in some schemes elsewhere, resentment can lead to sabotage of pipes and taps, jeopardising the work done by the operator.

At the same time, the local government level stakeholders, including the various interest groups, need to have a realistic understanding of what the proposed water and sanitation options and service delivery models mean, to avoid unrealistic expectations: what can truly be delivered, what are the technical and financial risks involved, what are the maintenance needs, and what are the other critical issues that may or may not appear later on? Water demand and wastewater management, sanitation, productive uses of water and sustainable offtake, pro-poor water governance, and public health issues are all relevant (Rautanen, 2016). These were all raised during the initial capacity building by the project. Murgia WUSC has managed to communicate this to its customers.

During the first years of operation the internal armed conflict in Nepal was still active. Murgia WUSC decided early on that it had to stay impartial and neutral, not responding to political influence and not supporting any political party. Initially WUSC members were selected through open elections every three years, following the processes as practised in local elections in Nepal. On two occasions there was heavy interference from the political parties, and the WUSC decided that open elections were not serving the purpose. Since then the WUSC members have been elected during the General Assembly. Within this political context, there was also pressure for donations, which in Nepal could sometimes be collected in a rather violent fashion. The WUSC's decision on donations was clear: 'absolutely not to anyone at all'. These two decisions that were made early on, have contributed to keep the system both socially and financially functional as revenues have been invested for the benefit and improvement of the system and the community. As reported in September 2016, all political parties are now fully supportive to Murgia WUSC, even doing fund-raising for the new works without expecting any favours in return. In the highly politicised environment of Nepal, this is an important lesson for replication elsewhere, or at least to ponder upon. In other schemes, we have seen political appointees who do not bother to come to work, or 'donations' that drain the WUSC coffers.

Murgia offers an important lesson to other communities, that keeping the WUSC free of political pressure or community conflict can contribute to its sustainability.

Financial practices and performance – is this replicable?

Murgia was designed, planned and implemented within the local government context. Its investment funds were directed through the District Development Funds. The total scheme cost was estimated at NPR 17.5 million (about USD 240,000 in year 2001). Of this, the governments of Finland and Nepal provided support of NPR 7 million each (about USD 96,000 each in year 2001). Of the total cost, the two governments and the District Development Committee covered 81%, the Parroha VDC 6% and the users 27%. Half of the users' contribution was in kind. The Murgia WUSC raised upfront with cash contributions from the beneficiaries: NPR 3,000 from each household and NPR 5,000 from each public tap stand, the latter cost being shared by those households that used these tap stands. The Murgia WUSA staff manages these tap stands. This cash was put into the Murgia WUSA's bank account so that if the water supply system was not finalised, the WUSA could pay back the debt.

Each private water meter is read once per month. The WUSC is flexible in accepting both cash and in-kind contributions from the users. A clear stipulation of associated costs is known and shared with the users. For example, a new tap connection is NPR 14,500[2]. The community is subject to a monthly progressive water tariff based on consumption, ranging from 80 NPR for the first 10 m^3 to 100 NPR for consumptions above 50 m^3. Penalties, water meter servicing, plumber's services such as new connections or change of locations are also costed. Computerised billing software is in use. Consumers can pay in advance – an option that many find convenient as they can avoid the monthly payment routine and pay when they have cash available. Many households have a family member working outside Nepal, and periodically sending remittances. In addition, crop sales do not produce a steady income. The WUSC is now planning to start sending SMS messages to its users, showing their balance/due payment and the last payment date. They also have a vision for mobile banking, making the water tariff payment convenient to its customers. The Murgia WUSC makes it clear for the users what kinds of services are available, and what are the associated costs.

In September 2016, the WUSA reported an average income of NPR 800,000 per month, with an average operating expenditure of NPR 128,190. At the closing of the last fiscal year (July 16, 2016) they had

[2] USD:NPR rate was 107 as of 25.12.2016.

a balance of NPR 25 million (equal at that time to about USD 230,000) even after major capital maintenance expenditure and new construction (below) over the fiscal year. Murgia WUSA is exceptional in comparison to other similar schemes in terms of having invested in new infrastructure and capital maintenance from its own resources and for having a high revenue/expenditure balance. It has been able to respond to the big challenge of how to expand services at a pace that allows the town to keep up with population growth, and maintain financial stability. Besides coping with regular operating and minor maintenance works such as salaries, allowances and service changes over the last Nepali Fiscal Year, the following capital expenditures (one-time costs of constructing fixed assets) were carried out:

- Construction of a new, accessible sanitation block that features separate toilets for women and men and includes a shower facility.
- Construction of a new drainage system in Ward 3. Murgia WUSA has a separate contract with the Division of Water Supply and Sewerage for the system. Over the past year they completed 3,500 m, with 83 manholes. The drainage system includes a treatment system.
- Construction of a new (separate, smaller) overhead tank to serve Wards 2 and 3. This separate system will be managed by the same WUSC.

Under capital maintenance works (occasional) costs of renewing (replacing, rehabilitating, refurbishing, restoring) assets to ensure that services continue at the same level of performance that was first delivered: Murgia WUSA painted all structures within the compound and the overhead tank itself, including repairs on the structures themselves and carried out maintenance of borehole well development to clear out turbid water.

For the coming year the Murgia WUSA has planned major works, including paving its' compound access road, improving the customers' waiting area, a new deep borehole, and launching a SMS service to send the water tariff payment information (payments due and balance), possibly with instructions on how to pay through a mobile banking system. With the drainage and related treatment unit for drainage water, Murgia WUSA is now also moving towards sewerage connections. The WUSC actively promoted sanitation earlier, refusing to give a water connection to households without toilets. In this way, the locality was declared 'open defaecation free' fairly quickly.

The normal process with government-supported water schemes is that the community does not procure the materials, nor is responsible for construction, even if all expect some form of local contribution. Capacity building is limited. This lack of community ownership and capacity is one reason for the problems with functionality in water schemes in general in Nepal (White *et al.*, 2015). Murgia's experience is replicable if there is strong financial discipline and the funds are used in a transparent way. This practice is linked to the two above points, the organisational dynamics and social coherence within the WUSA/WUSC itself. Naturally there are risks if poorer households are unable to pay, but to date the management has been able to bring the whole community on board, with all households connected and paying their bills.

What kind of technical and capacity development support did Murgia WUSA receive in the beginning? How is this replicable?

This system was defined from its foundation differently than many other systems in Nepal. Therefore, looking at what type of support it received is not only important, but essential. RWSSSP III aimed to build self-reliance among communities (including Murgia) and mobilise local resources. The Project used a participatory step-by-step process and gender-sensitive approach to social mobilisation and scheme planning and implementation, clarifying the actions and responsibilities of various actors at each stage. The WUSC was established as the main managerial body at the first step. Community level audits increased transparency and trust of the community, and hence their willingness to pay water tariffs. This community management functioned due to the hands-on support provided by the project in the form of technical advisors (Finnish and Nepali) and support persons based at community level. They ensured quality scheme design and construction, but also facilitated the 'soft' elements of training, community participation and institutional development.

This level of ownership was unique in Nepal at the time, and highly appreciated by the communities, increasing transparency, direct participation and interest of both women and men, and leading to strong sense of ownership that was crucial for the long-term sustainability of the facilities. The Completion Report quoted a local government officer: '*the RWSSSP* [*is*] *unique in this sense. Usually in the projects there are recipients and providers. The RWSSSP special kind of modalities created the situation, where all stakeholders were the partners of the RWSSSP (…)*' (RWSSSP, 2005). An evaluation of Finnish country programme in Nepal noted '*Finland's bilateral water supply projects have supported the creation of local systems that attract community support, participation, maintenance and cost recovery. Such systems are rightly seen as best practices*' (Caldecott *et al.*, 2012).

The Murgia WUSC members commented on the valuable role technical staff played in developing community management capacity, and quickly resolving problems, technical and social alike. The project modality was instrumental in ensuring that Murgia WUSA got off to a good start, even if the project modality has been criticised as hindrance and a subversion of local systems by some authors (e.g. Tarp & Hjertholm, 2000; OECD/DAC, 2005). The project modality enabled local application of the key factors affecting post-construction sustainability – technical, financial, community and social, institutional and environmental factors (Bakalian & Wakeman, 2009; Rautanen, 2016). It is notable that the majority of the other schemes also constructed during RWSSSP, are also still functional (as seen in monitoring to date by Finland-Nepal cooperation). The key element appears to be the strong foundation of technical support and the step-by-step process. Table 3 shows how Murgia has been able to adapt to the small-town environment and changing role, following this good start.

Conclusions

> '*Construction is not as important as operation and maintenance.*'
> Murgia WUSA Chairperson
> Captain Moti Prasad Rana, 10.09.2016

Murgia WUSA was successful even after 10 years of operation when compared against 63 other small town water service providers in Nepal on several accounts. It provides a quality service at low cost, while carrying out capital maintenance and investment. From the institutional and financial points of view this case study concludes that transparency and accountability are key success factors. Consumers trust that the funds are used appropriately and there is a high willingness-to-pay. The managers (the WUSC itself), its staff and the consumers benefit as there seems to be neither funds nor water leaking out of this system. A sense of ownership has translated in Murgia into a sense of responsibility to pay the water tariff to ensure continued services and the responsibility for not vandalising or abusing the structures, as well as responsibility for reporting the service problems for timely repairs.

Table 3. Factors influencing sustainability of rural water schemes – and how Murgia has responded as it moves to a small-town environment.

Pre-project issues	Post-project issues	How Murgia has responded
Community participation	Community ownership; sense of ownership; definition of roles and responsibilities for system management; both for service authorities and service providers	Clear roles for community at several levels of management. Cluster-wise autonomous sub-committees responding and being accountable to their own communities, giving the WUSA a human face and bringing WUSC members closer to their organisation. Full time professional committed staff, while the voluntary Executive Committee still represents the community
Scheme financing, contributions; developing transparent and accountable practices	Tariff collection for both regular O&M; financing for capital maintenance expenditure and reinvestment/extension/ service level improvements; public audits and public hearings are continued to maintain transparency and accountability of the user committees to the users	Transparent procedures, regular audits, continuing financing of capital maintenance and reinvestments. Policy to avoid political involvement. Interest to improve financial tools (e.g. online banking, etc.) Regular monthly meetings, ensuring that all Executive Committee members stay aware of social, technical and financial performance of the system, and can respond in a timely manner to any reported problems
Demand-responsive approaches; Serving-the-unserved; inclusion	User satisfaction; customers willing to pay for the expected services; reliable services meet the demand; inclusive services ensuring water for all as per the right to water	Willingness to pay Improving service levels to meet demand Active targeting of disadvantaged groups
Empowerment and capacity strengthening	Capacity of water user committees and its individual members; rights holders can call for their rights from the duty bearers; duty bearers have capacity and willingness to respond	Murgia WUSC received significant capacity building. Acting as a duty bearer
Appropriate technical design, construction quality, location of the structures, related environmental considerations and works	Linkages to government and private sector service providers and other local government/district level processes	Scheme was well designed, allowing service extensions Employment of paid staff
Gender quality and Social Inclusion (GESI) on participation and benefits	GESI on capacity enhancement	Active participation of women in committees
Training (for planning and implementation)	Continued learning-by-doing, refreshers, peer-networks (e.g., among pump operators, village maintenance workers, new water users committee members)	WUSC members train new members; out-going chairperson will continue as WUSC member in support of the new chairperson

Source: Modified by Rautanen (2016) from Bakalian & Wakeman (2009, p. 9).

From the social and organisational points of view, this case study concludes that the Murgia WUSA has been successfully organised into various committees and task forces that allow it to function efficiently and effectively in a small-town context. It received technical support and facilitation at local level, right from the start, building management capacities. Incentives and motivation of individuals is shown by the value given from and within the community to holding positions in any of the subcommittees.

As is typical for emergent small towns, Murgia WUSA's service area itself has changed radically, and will continue to change. Its position as a municipality next to the market corridor and the Indian border will ensure continued growth. Livelihoods will continue to diversify, and the related service expectations will continue to grow. Murgia WUSC continues to be pro-active by building another overhead tank water supply system and a drainage system with wastewater treatment, keeping itself up-to-date. The future plans include mobile banking and SMS service for water tariff payment. Its technical solutions are achievable by other schemes if they continue building on 'soft' capacities. Perhaps the only risk evident is sustainability of water supply, with the high current use of domestic water, and the competing need for agricultural water. It would be valuable for Murgia to consider consumer education to avoid waste.

Murgia WUSC's success is rooted in good water governance principles: participation, responsiveness, financial transparency, accountability and overall strong commitment and vision – all based on the initial step-by-step approach. We cannot claim these factors can be easily replicated, but they indicate that these factors are key to sustainability. These have kept the WUSA investing both in capital maintenance expenditure as well as new infrastructure. The findings from the Murgia WUSA resonate well with Ravet & Braïlowsky (2014), who call for empowerment and dialogue with all relevant stakeholders, even in the implementation and operation of large, clearly urban water utilities. They conclude that soft engineering skills have proven their efficiency, identifying more profound locality-specific social issues at the root of many problems, rather than technology or financing itself. It is of utmost importance for a WUSC to establish its non-negotiable principles and show accountability and transparency to the public from the very beginning. At least, this proved to be crucial in the case of Murgia WUSA. This becomes more critical as small towns grow and their perhaps originally rather homogeneous community becomes more complex. In order to guarantee the future of community-led small town water systems, we advocate for maintaining the human face of the providers.

In the small towns' context, the non-governmental water service providers' capacity and attitudes become critical if and when the government shifts away from its service provider role towards service authority. One static model is unlikely to work in a typical small towns and emerging towns' context, given the dynamic growth and change in these locations.

These findings are relevant for programme and policy-level debates both in Nepal and globally, the challenge being *how to have more of these types of successful water service providers in Nepal*? One element is capacity development and technical support during the early stages, setting the 'rules of the game' and facilitating a strong service culture. The enabling environment keeps changing, with the entire local governance structure being reconstructed as federalism proceeds in Nepal. The position of small towns is being defined in more detail in Nepal. This is a chance for Nepal to address the regulatory and technical standards related to consumers' associations and users' groups. The real life operational experiences from cases such as Murgia WUSA are most valuable now, when the new policies and regulations are being defined. Programmatic approaches to small town water services should consider sub national-level institutional arrangements, and give room for local adaptation and initiative.

Acknowledgement

The authors are grateful to the Murgia WUSC for their commitment and vision to provide the best water supply service for their community. This study was specially supported by the WUSC chairperson Mr. Moti Prasad Rana and the Office Manager Mr. Dinesh Acharya. This study was part of the Rural Water Supply and Sanitation Project in Western Phase II (Completion Phase) effort to document the lessons learned in the earlier Rural Water Supply and Sanitation Support Programme Phase III (1999–2005). Both projects are bilateral water and sanitation projects supported by the Governments of Nepal and Finland.

References

Bakalian, A. & Wakeman, W. (eds) (2009). *Post-Construction Support and Sustainability in Community-Managed Rural Water Supply – Case Studies in Peru, Bolivia and Ghana.* Water Sector Board Discussion Paper Series, paper 14. World Bank/Netherlands Water Partnership, Washington, DC, USA. http://documents.worldbank.org/curated/en/2009/06/10626231/post-construction-support-sustainability-community-managed-rural-water-supply-case-studies-peru-bolivia-ghana (accessed December 25 2016).

Caldecott, J., Hawkes, M., Bajracharya, B. & Lounela, A. (2012). *Evaluation: The Country Programme Between Finland and Nepal.* Evaluation Report 2012:2. Ministry for Foreign Affairs of Finland, Helsinki.

Flyvberg, B. (2006). Five misunderstandings about case-study research. *Qualitative Inquiry 12*(2), 219–245.

Government of Nepal (2001). *Nepal – National Population Census 2001, Tenth Census.* Central Bureau of Statistics, National Planning Commission Secretariat, Kathmandu, Nepal. http://cbs.gov.np/nada/index.php/catalog/42 (accessed November 25 2016).

Government of Nepal (2009). *National Urban Water Supply and Sanitation Sector Policy -Final Draft April 2009.* Ministry of Physical Planning and Works, Kathmandu, Nepal. http://dwss.gov.np/file/file_down/S4Io4uNational%20Urban%20Water%20Supply%20and%20Sanitation%20Sector%20Policy,%202009,%20Final%20Draft.pdf (accessed September 10 2016).

Government of Nepal (2011). *Water Supply, Sanitation and Hygiene – Sector Status Report, May 2011.* Ministry of Physical Planning and Works, Water Supply and Sanitation Division, Sector Efficiency Improvement Unit, Kathmandu, Nepal. http://www.seiu.gov.np/index.php/documents?folder=Sector%2BReports (accessed September 10 2016).

Government of Nepal (2014a). *Nationwide Coverage and Functionality Status of Water Supply and Sanitation in Nepal.* Ministry of Urban Development, Department of Water Supply and Sewerage, National Management Information Project (NMIP), Kathmandu, Nepal. http://www.seiu.gov.np/index.php/documents?folder=Sector%2BReports (accessed September 10 2016).

Government of Nepal (2014b). *National Water Supply and Sanitation Sector Policy 2014 – Final Draft December 31, 2014.* Ministry of Urban Development, Kathmandu, Nepal. http://dwss.gov.np/file/file_down/S4Io4uNational%20Urban%20Water%20Supply%20and%20Sanitation%20Sector%20Policy,%202009,%20Final%20Draft.pdf (accessed September 10 2016).

Government of Nepal (2015). *Water Service Providers Data Book, 2070–2071 (2013–2014).* Ministry of Urban Development in partnership with Second Small Towns Water Supply and Sanitation Sector Project (ABD project No. 41022/SEIUG3-NEP) and Asian Development Bank, Sector Efficiency Improvement Unit, Kathmandu, Nepal. http://www.seiu.gov.np/index.php/documents?folder=Publications (accessed September 10 2016).

Government of Nepal (2016). *Nepal Water Supply, Sanitation and Hygiene Sector Development Plan (2016–2030).* Ministry of Water Supply and Sanitation, Sector Efficiency Improvement Unit (SEIU), Kathmandu, Nepal. http://www.seiu.gov.np/index.php/documents (accessed September 10, 2016).

Lockwood, H. & Smits, S. (2011). *Supporting Rural Water Supply: Moving Towards A Service Delivery Approach.* Practical Action Publishing, Bourton on Dunsmore, Rugby, Warwickshire, UK. http://www.ircwash.org/resources/supporting-rural-water-supply-moving-towards-service-delivery-approach (accessed December 25 2016).

Moriarty, P., Patricot, G., Bastemeijer, T., Smet, J. & van der Voorden, C. (2002). *Between Rural and Urban: Towards Sustainable Management of Water Supply Systems in Small Towns in Africa.* International Water and Sanitation Center. January 2002. The Hague, The Netherlands.

Organisation for Economic Co-operation and Development/Development Assistance Committee (OECD/DAC) (2005). *The Paris Declaration on Aid Effectiveness*. Paris, 2005. https://www.oecd.org/dac/effectiveness/parisdeclarationandaccraagendaforaction.htm (accessed December 27 2016).

Rautanen, S.-L. (2016). *Access to Water? Dynamic Capacity Change for Sustainable Rural Water and Sanitation Services for All*. Tampere University of Technology, Publications 1373, Tampere, Finland. http://URN.fi/URN:ISBN978-952-15-3707-3.

Ravet, S. & Braïlowsky, A. (2014). Utilities' contribution to the human right to water and sanitation – importance of stakeholders' ownership. *Aquatic Procedia 2*, 70–78. doi: 10.1016/j.aqpro.2014.07.011.

Rural Water Supply and Sanitation Support Programme, Nepal, Phase III (RWSSSP III) (2005). *Completion Report*. Kathmandu. December 2005.

Sugden, F. (2014). *Landlordism, Tenants and the Groundwater Sector: Lessons From Tarai-Madhesh, Nepal. International Water Management Institute* (IWMI), Colombo, Sri Lanka, 33 pp. (IWMI Research Report 162). doi: 10.5337/2015.204.

Tarp, F. & Hjertholm, P. (2000). *Foreign Aid and Development: Lessons Learnt and Directions for the Future,* Routledge, London and New York.

White, P., Badu, I. R. & Shrestha, P. (2015). Achieving sustainable water supply through better institutions, design innovations and water safety plans – the experience from far west Nepal. *Journal of Water, Sanitation and Hygiene for Development 5*(4), 625–631. doi: 10.2166/washdev.2015.002.

WHO/UNICEF (2015). *Joint Monitoring Programme for Water and Sanitation – Online Data Table*. http://www.wssinfo.org/data-estimates/tables/ (accessed November 25 2016).

WSP (2011). *The Hard Way to the High Road: Transition of Community-Based Water Groups to Professional Service Providers in Indonesia*. Water and Sanitation Program: Learning Note.

Public finance for water infrastructure development and its practical challenges for small towns

Elena Humphreys[a], Andrea van der Kerk[b] and Catarina Fonseca[b]

[a]*Corresponding author. IHE Delft Institute for Water Education, Westvest 7, Delft 2611AX, The Netherlands*
E-mail: elenahumphreys@gmail.com
[b]*IRC, Bezuidenhoutseweg 2, The Hague 2594 AV, The Netherlands*

Abstract

The small and fluctuating population, the economic characteristics and administrative capacity of small towns not only pose infrastructural challenges for providing services, but also limit the possibilities for generating local revenues for financing water infrastructure development and maintenance. This limited ability to generate local resources for water infrastructure is exacerbated by the way in which scarce public funds are allocated. A first concern is linked to an urban bias that characterizes allocation of funds by central governments. A second concerns the prioritization of other sectors by allocation decisions of local governments. These local governments often prioritize other sectors such as education, health and agriculture for the use of scarce local public resources. What this discussion highlights is that existing models used for financing water infrastructure development do not seem very applicable to the realities of small towns. Additional research and models are necessary to allow for solutions that are better tailored to these realities.

Keywords: Financing; Public finance; Small towns; Water infrastructure

1. Introduction

Although it has a demonstrated an important role in infrastructure development, public finance has largely gone unaddressed in the literature. Instead, focus on financing has largely linked to the public-private debates over the past 25 years and concentrated on the promise and limitations of private sector investments (Winpenny, 2003). In line with this trend, research on public finance has largely been in terms of its deficiencies and the need for alternatives. This emphasis on private and blended finance stands in stark contrast with a finding from the World Bank, which states that two-thirds of all infrastructure in Africa was financed from domestically generated public revenue, with external aid supplying the remaining third (Foster & Briceño-Garmendia, 2009; Hall & Lobina, 2012). Data from the 1990s corroborate this finding as they demonstrate that almost 70 percent of infrastructure investments were

doi: 10.2166/wp.2018.007

sourced from the domestic public sector (Winpenny, 2003). Moreover, public finance will likely continue to be of importance as there has been a demonstrated lack of interest from the private sector to invest in water infrastructure in developing countries (Hall & Lobina, 2009).

Public finance is distinctly different from private finance as it is able to cheaply finance projects that benefit the public good (Hall & Lobina, 2012). Public finance is able to do this as it is backed by stable tax revenues and is able to invest in poorer areas, which are unattractive to commercial lenders (Massarutto *et al.*, 2008). However, while the initial investment may be sourced from public finance, it must be clarified that public investment does not necessarily mean that the services provided will be managed by a public entity. These services provide social and economic benefits beyond the individual private benefits of connecting to water and sanitation services. Apart from public health benefits, it is estimated that for each US$1 worth of investment US$7 is generated in economic returns (OECD, 2011). As the social and economic benefits are spread over society rather than just the individual user, having these services as a focus of government investment would appear justified (Gunther & Fink, 2010). The nature of public finance and of water and sanitation services indicate that public revenue remains an important source of financing for this sector.

1.1. Sustainable Development Goal 6 and public finance

With the adoption of the United Nations' (UN's) document on Transforming Our World through the Sustainable Development Goals (SDG) (UNGA 2015), the SDG were introduced as successor of the Millennium Development Goals (MDG). Sustainable Development Goal 6, ensuring availability and sustainable management of water and sanitation for all, addresses perceived shortcomings of the MDG target by emphasising universal water services by 2030. Recent estimates of the global coverage of basic water services, however, remain far below the goals of the SDGs. For example, basic water services have been provided to 87 percent of people living in urban areas and 76 percent of people living in rural areas (Hutton & Varughese, 2016). However, these numbers reduce to 68 and 20 percent, respectively for those who have access to safely managed water (Hutton & Varughese, 2016). For sanitation, more progress is needed as 26 percent of those living in rural areas have access to safely managed sanitation and only 34 percent of people living in urban areas have access to safely managed sanitation (Hutton & Varughese, 2016).

However, for this ambitious target to be achieved a considerable financing gap needs to be bridged. There is an estimated US$1.7 trillion investment gap for water and sanitation (Kolker & Trémolet, 2016a). As public finance comprises two-thirds of the available finance for water infrastructure, it is clear that the achievement of SDG 6 can only be achieved through significant public investment. This extremely large sum of money will likely be financed by the users, taxpayers, or from external aid (Winpenny, 2003).

1.2. Small towns

As global attention moves from the UN's MDGs to achieving the target of the SDGs, small towns must receive increased attention if universal access to water and sanitation is to be achieved (Habitat, 2006). Over the past decades, literature has mainly focused on water and sanitation in either urban or rural settings and largely ignored water services in small towns (WSP, 2002; Pilgrim, 2007). Small towns require specific attention because they straddle both urban and rural characteristics. Therefore research focusing solely on either rural or urban contexts may create recommendations inappropriate for small towns (Caplan & Harvey, 2010). In addition, small towns are of increasing importance as the number of small towns is growing due to increased urbanization. In fact, it has been estimated that 20–40 percent of the population in Africa and Asia lives in this grey area between urban and rural settings (Pilgrim, 2007).

Despite the importance of public funding in water supply and sanitation and the increasing prominence of small towns, both topics have received little attention in recent years. This article, which is based on a thorough review of existing (grey) literature, contributes to filling this gap. Public finance is used for both the development of infrastructure and on-going operations and maintenance. However, this paper primarily focuses on its role in infrastructure development. In doing so, this paper seeks to clarify the role of public finance through a review of the current sources of generation for public revenue, the different allocative theories, and the associated lines of accountability. This article then contextualizes the challenges of public finance for both infrastructure development in reference to small towns. Accordingly, this article highlights that while public finance plays an important role in the financing of water and sanitation infrastructure and should not be overlooked, there are significant challenges for its use in small towns.

2. Public finance

Public finance refers to the government's role in the economy, which acts to minimize undesirable effects of a market economy (market failure) and maximize desirable effects (Gruber, 2004). Public revenues are generated to finance public expenditures. There are two main sources of domestic public revenue: taxation and tariffs[1]. However these standard sources of revenue may be insufficient to generate the needed revenue to match the lump investment needed for infrastructure development. To fill this gap, borrowing instruments have been promoted to fund governmental services (OECD, 2009). While this paper focuses on sources of revenue channelled through the government, there are additional revenue flows from external transfers, not discussed in this article, that are provided on a 'project by project' basis.

2.1. Revenue generation: taxes and tariffs

At its most basic, taxation is the assessment of a levy by the government on an individual or legal entity (OECD, 2014). The main justifications for the levying of taxes are to generate revenues for public services, redistribute revenue for equity, internalize the externalities of the production of certain goods, and to ensure representation (Taxation and Finance, 2013). The revenues collected from these individuals or legal entities are then used to support general services for which the market will not provide (Hall & Lobina, 2009). Taxation comes in many forms; from sales taxes on the purchase of goods to the assessment of royalties on natural resource extraction. Taxes can be collected at the different levels of government, from local sales taxes to centrally collected income taxes (McLure Jr, 2001).

[1] The Camdessus Report on Financing of water infrastructure cites three main sources of funding for the water sector: taxation, tariffs and transfers. Transfers can be categorized as international public finance, while taxation and tariffs are both sources of domestic public finance (Winpenny, 2003).

The amount of revenue generated through taxation will depend on the size of the tax base. Therefore taxes collected at the central government level will provide a larger pool of money than those collected at the local level (Alm, 2015). Taxation is typically considered a relatively stable form of public revenue generation, as non-payment typically can lead to civil penalties (Taxation and Finance, 2013). However, to ensure proper enforcement, taxation systems require administrative capacity to manage the financial flows and properly assess taxes. Taxation typically makes up the significant portion of the public revenue generated for public services.

Tariffs are the other primary source of public revenue (Winpenny, 2003). In short, tariffs are payments for access to or use of services. This can be as simple as the use of a toll road or payment for water supply from a tap or piped network. These tariffs support the operations and maintenance of these services and can be used to cross-subsidize services (Winpenny, 2003). However as a source of public finance, tariff collection can be subject to instability. The instability is caused by conditions of uncertainty regarding the user base. If the user base decreases then tariffs may be insufficient to adequately fund the operations and maintenance of the service. Therefore while tariffs are an important source of public revenue they are likely less stable than revenues from taxation.

2.2. Loans

While revenue generated through taxation and tariffs is the foundation of public finance, they are typically insufficient to finance particular large capital expenditures needed for infrastructure works. Accordingly, governments at the central, subnational, and local level can take on debt to finance these capital expenditures. Governments will issue debt, when a large capital expenditure requires a lump investment upfront, but the benefits from the investment will be long lasting (Alm, 2015). These debt obligations attract investment from external and internal investors and are backed by promises of repayment from user fees issued on the infrastructure (USAID, 2003; Kolker & Trémolet, 2016b). Borrowing is based on the theory of intergenerational equity, which specifies that burden and risk of financing such projects should be shared by those who benefit now and those who benefit in the future (Martell, 2000).

Borrowing can be problematic as it poses a number of risks and challenges. For governments to access capital markets through the issuing of bonds, they must demonstrate a number of characteristics including credit worthiness, insurance against default, and other risk mitigation measures (Winpenny, 2003). However, these characteristics that enable national, subnational and local governments to access capital markets are often lacking, making it a challenge to attract investors (Cardone & Fonseca, 2006). When these characteristics are lacking, governments must attract investors by offering higher than average interest rates and other credit enhancements (Securities and Exchange Commission, 2012). These mitigations drive up the price of municipal debt.

2.3. Allocation: subsidiarity and distributive equity

Public revenue generated through these sources can be allocated consistent with different allocation principles of public finance. Broadly speaking, allocation happens in two ways. The first (subsidiarity principle) refers to the level of government that makes allocative decisions about the use of public revenue. The second (distributive equity) principle refers to specific allocative purposes for the use of the public finance.

The subsidiarity principle states that public service provision should be the responsibility of the lowest level of government that can do so effectively. This principle assumes that the local government is in the best position to understand the needs of their constituency and therefore will provide services efficiently by matching the supply of services with the demand for services. While this theory primarily addresses the management of public services, it has also been applied to infrastructure financing (Alm, 2015). As local governments are best positioned to understand the infrastructure needs of their populations, they are also in the best position to distribute the costs according to who benefits from the development of infrastructure. This is echoed by McLure Jr (2001), who argues that local governments are often better able to make tax decisions than central governments as they are closer to the people and therefore can efficiently provide government services. Regardless of the source of public revenue, allocation decisions are made with relative freedom by the local government as they are not constrained by earmarks or conditionalities by regional or central government authorities. Accordingly for water infrastructure, local governments have autonomy to determine the location and type of water infrastructure investments.

The principle of distributive equity is informed by the priorities and targets national governments set for certain sectors. In order to ensure that these priorities are realized the government attributes financial resources for their achievement (Aboma, 2010). The distribution of this public revenue, thus, is accompanied by strict requirements that govern the use of the funds. Under this principle central allocation seeks to address horizontal imbalances through redistribution. These horizontal imbalances are created when local governments differ in their ability to raise local revenue. By maintaining minimum funding levels for the nationally established priority areas, the central government ensures that the local governments have access to sufficient resources, which they are required to spend on specified activities.

2.4. Upward and downward accountability

Similarly to how private finance must respond to their shareholders and investors about the use of funds, the institutions that use public finance are accountable for their decisions. However the direction of accountability of public funding is also informed by the allocative theories that govern its use. These theories shape the pathways of accountability and can steer the use of public finance towards different purposes. While revenue generation can happen through taxation, tariff collection, and borrowing, the direction of accountability is often defined by the allocative theory for the use of public finance.

Under the first allocative theory of subsidiarity, accountability is downward; linked to the constituents that are receiving the services. As the decisions of allocation under this theory are made at the lowest possible level, accountability is oriented towards fulfilling the needs of the constituency. This is primarily because these constituents are, in theory, able to hold their elected officials accountable for their actions therefore they are incentivized to make decisions that best serve their constituents (Tullock, 2005).

In contrast, in systems where allocation of public revenue mirrors distributional equity, accountability on the use of public revenue focuses upward. As stipulations and earmarks are determined by the national government, the recipients of public revenue are then required to report the use of these funds according to the priorities (Aboma, 2010). Therefore the monitoring and enforcement of the use of public revenue are then focused on the national government (Mehta & Mehta, 2008). Accountability helps determine the development of infrastructure as it develops according to the national priorities rather than based on the input of the constituency.

3. The challenge of water infrastructure development in small towns

Small towns are particularly challenged by infrastructure development and maintenance because of their characteristic low density and small, but fluctuating populations. Franceys *et al.* (2016: 84) attribute challenges of water service provision in small towns to 'higher costs (reduced economies of scale, more expensive access to professional support) and lower resources/willingness to pay due to lower average incomes as well as better alternative water sources than in larger towns'.

3.1. Population size and water infrastructure development

The most frequently used characteristics to describe small towns is population size (Caplan & Harvey, 2010). Typically, the population size considered to define small towns can be as low as 2,000 to 20,000 people (Habitat, 2006), or as high as 5,000 to 50,000 people (Ndaw, 2016). The discrepancy in these population ranges demonstrates the challenges of defining small towns, as a small town in India is likely to be much larger than a small town in Ethiopia (Caplan & Harvey, 2010). Despite the different understandings of population size as a defining factor for small towns, it remains important as small towns typically cannot benefit from the economies of scale unless they reach a certain size (Ndaw, 2016). Economies of scale refers to the concept that the unit costs of service provision reduces as the user base increases[2]. Population size is then an important characteristic for small towns since it determines the user base over which service provisioning costs can be spread (Habitat, 2006). While population size is important for determining economies of scale, the contestations over the exact population that makes a small town has led to criticism of only using population size to define small towns (Owusu, 2005). Accordingly, some governments have included characteristics such as density or economic activities as well as population to define small towns (Caplan & Harvey, 2010). However, the population size of small towns is important, as it affects the types of infrastructure available in the small town.

3.2. Population density and water infrastructure development

Another characteristic of small towns that affects infrastructure development is their low population density (Caplan & Harvey, 2010). Typically the development of a small town begins around the commercial centre (Hopkins, 2003). In the areas surrounding the centre, settlements develop with reducing density on the fringes. The development of infrastructure typically follows a similar pattern with the population density (Adank *et al.*, 2016). The centre of the towns are typically served by piped infrastructure. However, given the low density on the outskirts of town, piped service is often not expanded (Caplan & Harvey, 2010). Accordingly the small towns are also serviced by standpipe infrastructure (Adank *et al.*, 2015). In their spatial distribution, small towns mimic both urban and rural characteristics. Some authors have linked the low population density to the lower price of land on the fringes of small towns (Ndaw, 2016). However the low population density (but larger population than rural areas) creates

[2] Economies of scale tend to exist where production requires high fixed costs relative to variable costs, so that average costs for service provision decline continuously over the relevant range of output.

a number of challenges for service provisioning in small towns, as uniform infrastructure is often not economically viable.

3.3. Population fluctuation and water infrastructure development

Related to the density characteristics of small towns, the population of small towns are typically in flux. They are subject not only to periods of immigration, but also population reductions due to emigration (Caplan & Harvey, 2010). Some authors have focused on the growth of small towns as being the centres for urbanization (Adank *et al.*, 2015), while others state that small towns are pass-throughs for migration to larger urban centres (Trager, 1988). The causes of rapid population growth in small towns have been attributed to both natural disaster events and to policy decisions (Caplan & Harvey, 2010). Emigration from small towns has been attributed to the limited economic activities that exist in such towns (Owusu, 2005). Regardless of the cause, these fluctuations challenge infrastructure development in small towns as the nature of water infrastructure development is that it is somewhat permanent. Incorporating flexibly to meet these population changes is then a significant challenge.

4. Discussion: the challenge of public finance of water infrastructure in small towns

Small towns have small populations, which means that their revenue generation capacity is limited by the size of their tax base. In addition, the mix of both rural and urban economic activities (Caplan & Harvey, 2010) further limits their revenue generation capacity. While the exact mix of activities varies from small town to small town, the common economic activities in a small town are often light industry, trade and agricultural activities (Caplan & Harvey, 2010). Often, farmers sell their local agricultural commodities, sometimes at wholesale, to traders (Trager, 1988). These traders and farmers create the linkages between urban and rural areas, creating outlets for goods and agricultural commodities. This contrasts with more rural economies where the primary activity is agriculture and with urban areas where the economic activities can be classified as trade and industry based. Accordingly, the economy in small towns is both cash based and self-provisioning (Trager, 1988; Hopkins, 2003). The mixture of both trade, industry, and agricultural economic activities means that local government faces challenges in collecting taxes on the sale of goods in the marketplace (Englebert & Sangare, 2010).

These economic activities also inform the household income characteristics in small towns, as some have salaried work, while others rely on informal activities like day-labour or agricultural self-provisioning to sustain themselves (Adank *et al.*, 2015). In addition, due to their linkage to urban areas, households in small towns sometimes rely on remittances from relatives working outside the small town (Trager, 1988). The heterogeneity of the sources of household income leads to challenges in local revenue generation for several reasons. First of all, many households are unlikely to have a steady and reliable source of income (Caplan & Harvey, 2010), which create challenges for tax collection as taxation relies on regular administration of taxes. These irregular incomes also affect the ability to pay user fees for basic services (Caplan & Harvey, 2010; Franceys *et al.*, 2016). If residents are subject to a monthly water tariff for water and sanitation services, they will be constrained to pay these fees as it would take financial management to ensure that they could meet the schedule of tariffs. As small towns already have a smaller tax base, it is clear that typical systems of taxation and tariff collection are highly challenging for small towns. Second, despite remittances, small towns are defined by their high rates of poverty, attributed to the fact that a portion of the population is self-provisioning and that there is a small labour market (Ndaw, 2016). A study in India found that nearly one-fifth of residents in a sampling of small towns were living below the poverty line (Tiwari, 2016). Similarly, a study of small towns in Ethiopia found that women were the head of the households in approximately 24–48% of households. This contributes to low revenue generation, as female-headed households frequently have more incomes below the poverty line than in male-headed households (Adank *et al.*, 2015).

Apart from the challenges of raising local revenues, many local governments prioritize other sectors for the use of municipal resources over the water services sector. A WaterAid study examining the development of water infrastructure in Sub-Saharan Africa raises concerns about 'the effective use or targeting of the scarce resources that are available to the sector' (WaterAid, 2014). The study suggests that 'in practice districts prioritise other sectors, in particular health, education and agriculture'.

4.1. The limited capacity to access other sources of public finance

The limited capacity to locally generate sufficient public finance to invest in water infrastructure means that small towns depend on central sources of public finance or borrowing. However, also accessing these sources of public financing proves challenging for small towns.

4.1.1. The variability of centrally raised public finance. Despite the larger tax base of centrally generated public revenue, centrally raised public funds for water infrastructure are still subject to variations and periodic reductions. These varying or decreasing levels of funding lead to a piecemeal approach in water and sanitation services in small towns, as small towns are required to meet the funding stipulations and maintain their existing services with fluctuating levels of funding. An example from Uganda illustrates the variation in the level of water and sanitation funding, which varies from year to year. Central government funding has decreased from 7 percent of gross domestic product in 2004/05 to 2½ percent in 2011/12. In addition to these budget cuts, the depreciation of Ugandan currency has also contributed to the further reductions in available funding for water and sanitation infrastructure (Matyama, 2012). Variation of revenue generation are also likely to occur on locally generated sources of public finance due to their dynamic economic characteristics.

Small towns often struggle to manage the varying schedules and varying amounts of the different sources of revenue. In addition to the management of several funds, the transfers of centralized sources of revenue are sometimes delayed, which creates instability for small towns. For example, in Ghana, small towns receive support from the District Assemblies Common Fund, which is a centrally generated revenue source. This fund has been subject to a number of management challenges as the disbursements to the district assemblies are often delayed for up to a year (Abbey *et al.*, 2010). These delays are especially problematic in small towns due to their limited revenue generation, unlike larger urban areas, which are able to redirect their locally generated revenue to augment these transfers. Accordingly, the use of centralized public finance tends to favour recurrent expenditures like salaries rather than capital expenditures for infrastructure development (Mehta & Mehta, 2008). Others, however, find that, in situations where local governments lack the capacity to spend their budgeted resources and risk being subject to budget cuts from the central government, they often rush to expend the remaining revenues on unsustainable or poorly planned projects (Das *et al.*, 2016). In addition, without adequate accountability measures, local allocative decisions can favour a select few rather than the needs of the public (Das *et al.*, 2016).

4.1.2. The limited capacity to access capital markets. Small towns have been argued to have low administrative capacity, which further limits their ability to use public finance (Cardone & Fonseca, 2006; Ndaw, 2016). This limited administrative capacity is described in relation to access to and the management of borrowing instruments (OECD, 2009). Typically, debt financing with outside investment requires that investor confidence is guaranteed through state-backing, credit enhancements, or otherwise high interest rates (Peterson, 1998; Securities and Exchange Commission, 2012). To access capital markets, municipalities must guarantee the repayment of debt through effective and complete user fees or tax collection (Peterson, 1998). While it is clear that small towns are challenged even with collecting sufficient revenue, they also lack sufficient financial management capacities to provide investor confidence. Illustrative of the limitations of administrative capacity are the low levels of local repayment for subnational pooled financing vehicles. In Ghana, repayment was about 32% for a revolving fund administered by the Association of Water and Sanitation Development Boards (Agbenorheri & Fonseca, 2005). In Burkina Faso, repayment rates were low at about 20 percent but were improved to 80 percent through an awareness campaign (Cardone & Fonseca, 2006). Given that the rate base is small and that the populations are in flux, small towns typically do not demonstrate the necessary creditworthiness to access subnational pooled finance vehicles or to issue municipal debt without significant external support (USAID, 2003; Platz, 2009; Kolker & Trémolet, 2016b). It is also unlikely that they have the necessary accounting and financial management capacities to meet the obligations of debt service.

4.2. The limited influence of small towns on national agendas

When public finance is allocated to achieve distributional equity, small towns are challenged to engage with the national government to ensure that their needs are captured in these nationally established funding priorities. Small towns are unlikely to be able to maintain a lobby presence at the national level, which means that their funding priorities and interests may be overlooked. This is reflected by Gugler (1997) who finds that governments (re)allocate resources to achieve three different goals: (1) to improve the environment of decision-makers (placating behaviour); (2) to assure the continued collaboration of the middle class; and (3) to placate strategic elements of labour. This results 'in public resources being spent disproportionately on the privileged consumption of the few and conspicuous investment for the few in cities' (Gugler, 1997; Bakker, 2003).

As small towns are constrained in their ability to register their funding needs at the national level, the allocative decisions may not reflect the context of small towns or the needs of the inhabitants of small towns. Illustrative is an example from Uganda, where funding decisions for water and sanitation are primarily made in conjunction with the development of a national sector investment plan (WSP, 2011). These plans have strongly prioritized the development of infrastructure to the detriment of the maintenance of existing infrastructure, which leads to the waste of public resources (Matyama, 2012). This waste is demonstrated by an anecdote, which described the new construction of a water tank only 20 meters from an existing tank (KOIS Development Consultants, 2011).

5. Conclusion

The challenge for public financing of development of water infrastructure in small towns consists of a number of interrelated dimensions. The relatively small and fluctuating population, the economic characteristics and administrative capacity not only pose infrastructural challenges for providing services, but also limit the possibilities for generating local revenues for financing water infrastructure development and maintenance. At the same time, two biases that characterize most countries lead to the limited availability of scarce public funds for water infrastructure in small towns. The first concern is linked to the distributive equity principle. Under this principle central allocation seeks to address horizontal imbalances through redistribution. However, central governments may not necessarily have a desire to strive for the reduction of horizontal imbalances. The urban bias discussed above suggests that these horizontal imbalances may not be reduced, as central government prioritize certain geographical spaces and regions over others.

The second concern is more related to the subsidiarity principle and focuses on the ability of local government to better match supply and demand for services. Existing research on local government expenditure in small towns, however, in principle suggests that local governments often prioritize other sectors such as education, health and agriculture for the use of scarce local public resources. The demand for other services is thus prioritized over the demand for water services. The principle of subsidiarity then does not necessarily lead to additional funding for water infrastructure.

The challenges of public financing do not lead to easy solutions. Although commercial financing is increasingly being promoted as a way to address the financing gap in the water services sector, the question remains to what extent such commercial financing is really suitable for small towns. Without external support, few small towns are likely to be able to access commercial financing. What this discussion highlights is that existing models used for financing water infrastructure development do not seem very applicable to the realities of small towns. Additional research and models are necessary to allow for solutions that are better tailored to these realities.

References

Abbey, C., Azeem, V. A. & Kuupiel, C. B. (2010). *Tracking the Ghana District Assemblies Common Fund.* https://namati.org/resources/tracking-the-ghana-district-assemblies-common-fund/ (Accessed July 15, 2017).

Aboma, G. (2010). *Think Local, Act Local II Impediments to Effective Financing of Sanitation Services in Ethiopia: The Case of Three Local Governments.* WaterAid, London.

Adank, M., Butterworth, J., Godfrey, S. & Abera, M. (2016). Looking beyond headline indicators: water and sanitation services in small towns in Ethiopia. *Journal of Water Sanitation and Hygiene for Development* 6(3), 435–446.

Adank, M., Dickinson, N., Butterworth, J., Mekonta, L., Abera, M. & Bostoen, K. (2015). *WASH Services in Small Towns: Baseline Report for A Quasi-Randomised Control Trial to Assess Impacts of the One WaSH Plus Programme. IRC* (One WASH Plus Programme Report). Retrieved from http://www.ircwash.org/sites/default/files/wash_services_in_small_towns_baseline_report_final_13042015.pdf.

Agbenorheri, M. & Fonseca, C. (2005). *Local Financing Mechanisms for Water Supply.* https://www.ircwash.org/sites/default/files/Agbenorheri-2005-Local.pdf (Accessed August 5, 2017).

Alm, J. (2015). Financing urban infrastructure: knowns, unknowns, and a way forward. *Journal of Economic Surveys 29*(2), 230–262.

Bakker, K. (2003). Archipelagos and networks: urbanization and water privatization in the South. *The Geographical Journal 169*(4), 328–341. Retrieved from http://www.jstor.org/stable/3451572 (Accessed June 20, 2017).

Caplan, K. & Harvey, E. (2010). *Small Town Water and Sanitation Delivery: Taking a Wider View.* WaterAid, London.

Cardone, R. & Fonseca, C. (2006). *Experiences with Innovative Financing: Small Town Water Supply and Sanitation Service Delivery*. https://www.ircwash.org/sites/default/files/Cardone-2006-Experiences.pdf (Accessed August 20, 2017).

Das, B., Fernández Fernández, C., van der Gaag, N., McIntyre, P. & Rychlewski, M. (2016). *Water Integrity Global Outlook*. http://www.waterintegritynetwork.net/wigo/ (Accessed March 18, 2017).

Englebert, P. & Sangare, N. (2010). *Comparative Assessment of Decentralization in Africa: Burkina Faso Desk Study*. Retrieved from http://pdf.usaid.gov/pdf_docs/PNADX214.pdf (Accessed August 10, 2017).

Foster, V. & Briceño-Garmendia, C. (2009). *Africa Infrastructure Country Diagnostic*. The World Bank, Washington, DC.

Franceys, R., Cavill, S. & Trevet, A. (2016), Who really pays? A critical overview of the practices of funding universal access. *Waterlines 35*(1), 78–93.

Gruber, J. (2004). *Public Finance and Public Policy*. Macmillan, Basingstoke.

Gugler, J. (1997). *Cities in the Developing World: Issues, Theory, and Policy*. Oxford University Press, Oxford.

Gunther, I. & Fink, G. (2010). Water, sanitation and children's health: evidence from 172 DHS surveys (English). Policy Research working paper, no. WPS 5275. Retrieved from http://documents.worldbank.org/curated/en/630921468151763344/Water-sanitation-and-childrens-health-evidence-from-172-DHS-surveys.

Habitat (2006). *Meeting Development Goals in Small Urban Centres. Water and Sanitation in the World's Cities 2006*. Earthscan, London.

Hall, D. & Lobina, E. (2009). *Paying for Water and Sanitation the Essential Role of Public Finance*. http://www.right2water.eu/sites/water/files/2009-03-W-finance.pdf (Accessed September, 2017).

Hall, D. & Lobina, E. (2012). *Financing Water and Sanitation: Public Realities*. http://www.world-psi.org/sites/default/files/documents/research/psiru_financing_water_sanitation.pdf (Accessed June 11, 2017).

Hopkins, R. (2003). *An Alternative Perspective on WSS Services: Towns and the Urban/Rural Divide*. Town water supply and sanitation: companion papers, volume 3, Bank-Netherlands Water Partnership Project 043. World Bank, Washington, DC.

Hutton, G. & Varughese, M. (2016). *The Costs of Meeting the 2030 Sustainable Development Goal Targets on Drinking Water, Sanitation, and Hygiene*. http://www.worldbank.org/en/topic/water/publication/the-costs-of-meeting-the-2030-sustainable-development-goal-targets-on-drinking-water-sanitation-and-hygiene (Accessed July 10, 2017).

KOIS Development Consultants (2011). *A Study on Working with Parliament Towards Improving WASH Governance in Uganda*. WaterAid, London.

Kolker, J. & Trémolet, S. (2016a). *Achieving Universal Access to Water and Sanitation by 2030 – How Can Blended Finance Help?* World Bank, Washington, DC. Retrieved from http://blogs.worldbank.org/water/achieving-universal-access-water-and-sanitation-2030-how-can-blended-finance-help?CID=WAT_TT_Water_EN_EXT (Accessed October 3, 2016).

Kolker, J. & Trémolet, S. (2016b). *Case Studies in Blended Finance for Water and Sanitation*. World Bank, Washington, DC. Retrieved from http://wsp.org/sites/wsp.org/files/publications/WSS-9-Case-Studies-Blended-Finance.pdf.

Martell, C. R. (2000). *Transformation of Municipal Credit Systems in Developing Countries: Evidence From Brazilian Municipalities*. PhD School of Public and Environmental Affairs, Indiana University, Bloomington.

Massarutto, A., Paccagnan, V. & Linares, E. (2008). Private management and public finance in the Italian water industry: a marriage of convenience? *Water Resources Research 44*(12). doi:10.1029/2007WR006443.

Matyama, F. (2012). *Financing of the Water, Sanitation and Hygiene Sector in Uganda*. WaterAid, London. Retrieved from http://www.wateraid.org/ ~ /media/Files/UK/Keeping Promises case studies/Uganda the financing of the water sanitation and hygiene sector.pdf (Accessed March 10, 2017).

McLure Jr, C. E. (2001). The tax assignment problem: ruminations on how theory and practice depend on history. *National Tax Journal 54*(2), 339–363.

Mehta, M. & Mehta, D. (2008). *Financing Water and Sanitation at Local Levels: Synthesis Paper*. WaterAid, London.

Ndaw, M. F. (2016). *Private Sector Provision of Water Supply and Sanitation Services in Rural Areas and Small Towns*. http://documents.worldbank.org/curated/en/450101468179030315/pdf/104505-WPS-Box394877B-PUBLIC-Add-series-WSP.pdf (Accessed April 10, 2017).

OECD (2009). *Innovative Financing Mechanisms for the Water Sector*. OECD Publishing, Paris.

OECD (2011). *Benefits of Investing in Water and Sanitation: An OECD Perspective*. OECD Publishing, Paris. http://dx.doi.org/10.1787/9789264100817-en.

OECD (2014). *Fundamental Principles of Taxation*. OECD Publishing, Paris.

Owusu, G. (2005). Small towns in Ghana: justifications for their promotion under Ghana's decentralisation programme. *African Studies Quarterly 8*(2), 48–69.

Peterson, G. E. (1998). *Measuring Local Government Credit Risk and Improving Creditworthiness.* http://documents.worldbank.org/curated/en/559551468314072571/pdf/375550Local0go1credit0risk01PUBLIC1.pdf (Accessed April 23, 2017).

Pilgrim, N. (2007). *Water Working Notes: Principles of Town Water Supply and Sanitation, Part 1: Water Supply.* Water Supply and Sanitation Sector Board of the Infrastructure Network, World Bank, Washington, DC.

Platz, D. (2009). Infrastructure finance in developing countries: The potential of sub-sovereign bonds. *Department of Economic and Social Affairs Working Paper, 76.* http://www.un.org/esa/desa/papers/2009/wp76_2009.pdf (Accessed February 9, 2017).

Securities and Exchange Commission (2012). INVESTOR BULLETIN Municipal Bonds: Understanding Credit Risk. *Office of Investor Education and Advocacy* (134). Retrieved from https://www.sec.gov/investor/alerts/municipalbondsbulletin.pdf (Accessed April 12, 2017).

Taxation and Finance (2013). Retrieved from http://finance.mapsofworld.com/finance/taxation/ (Accessed April 29, 2017).

Tiwari, A. (2016). Is slipshod urban planning, hampering sustainability of small Indian towns? *Procedia – Social and Behavioral Sciences 216*, 9–1.

Trager, L. (1988). Rural-urban linkages: the role of small urban centers in Nigeria. *African Studies Review 31*(03), 29–38.

Tullock, G. (2005). *Bureaucracy: The Selected Works of Gordon Tullock,* Vol. 6, Liberty Fund Inc., Indianapolis.

UN General Assembly (UNGA) (2015). Transforming our world: the 2030 Agenda for Sustainable Development, 21 October 2015, A/RES/70/1, available at: http://www.refworld.org/docid/57b6e3e44.html.

USAID (2003). Pooled Finance Model for Water and Sanitation Projects: The Tamil Nadu Water and Sanitation Pooled Fund (WSPF). *USAID Policy Note, 31.* Retrieved from http://pdf.usaid.gov/pdf_docs/PNACT334.pdf (Accessed August 19, 2017).

WaterAid (2014). *From Promise to Reality: The Urgent Need for Southern African Leaders to Deliver on Their Water, Sanitation and Hygiene Commitments,* Retrieved from http://www.wateraid.org/what-we-do/our-approach/research-and-publications/view-publication?id=af60d21e-3493-4b6e-b30c-98633054ac24 (Accessed August 22, 2017).

Winpenny, J. (2003). *OECD Global Forum on Sustainable Development: Financing Water and Environmental Infrastructure for all.* Michel Camdessus Panel, Paris.

WSP (2002). *Small Towns, Special Challenge.* World Bank, Washington, DC. Retrieved from http://www.wsp.org/sites/wsp.org/files/publications/small_towns.pdf (Accessed April 29, 2017).

WSP (2011). *Water Supply and Sanitation in Uganda: Turning Finance Into Services for 2015 and Beyond.* World Bank, Washington, DC. Retrieved from https://wsp.org/sites/wsp.org/files/publications/CSO-uganda.pdf (Accessed March 12, 2017).

IWA Publishing's authorised EU representative for General Product Safety Regulations is Diane D'Arras, 15 rue Duret, 75116 Paris, France, e-mail: safety@iwap.co.uk.

Printed and bound by CPI Group (UK) Ltd, Croydon, CR0 4YY
12/05/2026
02108891-0020